THE COSMIC INQUIRERS

The
COSMIC INQUIRERS

—— *Modern Telescopes and Their Makers*

—— WALLACE TUCKER
KAREN TUCKER

HARVARD UNIVERSITY PRESS
Cambridge, Massachusetts
and London, England 1986

*This book is printed on acid-free paper, and its binding materials
have been chosen for strength and durability.*

Library of Congress Cataloging-in-Publication Data

Tucker, Wallace.
    The cosmic inquirers.

    Bibliography: p.
    Includes index.
    1. Telescope.   2. Space telescope.   3. Astronomy.
4. Astronomers.   I. Tucker, Karen.   II. Title.
QB88.T83   1986   522'.2'0904   85-21952
ISBN 0-674-17435-6 (alk. paper)

For Charles, Josephine, Kerry, and Stuart

___ ACKNOWLEDGMENTS

Since this book was primarily undertaken as an effort to introduce the reader to some of America's foremost astronomers as well as to give an overview of the state of modern astronomy, our ability to obtain interviews was crucial. We found that this portion of the project was the most enjoyable. We were always courteously received; our interviews and discussions with these scientists were, without exception, pleasurable and informative. We thank John Bahcall, Carl Bignell, Barry Clark, Carl Fichtel, Gerald Fishman, Riccardo Giacconi, Paul Gorenstein, David Heeschen, Allan Jacobson, James Kurfess, Frank Low, Gerry Neugebauer, C. R. O'Dell, B. T. Soifer, Lyman Spitzer, Jr., and Harvey Tananbaum for their time and candor. We are also grateful to Hale Bradt, Harvey Tananbaum, and Patrick Henry for helpful comments and suggestions. In addition, we thank Angela von der Lippe and Susan Wallace for their editorial encouragement and help.

___ CONTENTS

Illustrations follow page 124

THE COSMIC INQUIRERS

── *PROLOGUE*

Ever since some long-forgotten stone-age Einstein designed Stone-henge, man has been making and using tools to investigate the universe and to try to understand its workings. These tools, like all tools, are in essence extensions of human faculties. Since astronomy is the science that attempts to observe and describe the universe as a whole, it is not surprising that astronomical observatories are where the grandest extensions of man's faculties are to be found. There is the Big Eye on Palomar Mountain, the Big Ears, as the giant radio telescopes have been called, the long, long arms of the *Viking* Lander missions to Mars, and the x-ray vision of the *Einstein* x-ray telescope. Modern telescopes, and all that goes into making them as well as the astronomical information gathered from them, have an importance that transcends science. They have had and will continue to have a profound intellectual impact on us, our culture, and our history.

Telescopes are used to look at objects that are incomprehensibly far away in terms of our normal experience. Perhaps some idea of their capability can be had by considering what they could detect at the distance of the moon, which is roughly a quarter of a million miles: The 200-inch optical telescope on Mt. Palomar could detect a ten-watt light bulb, the Space Telescope could detect a one-quarter-watt light bulb, the Very Large Array radio telescope could detect the signal from a small walkie-talkie, the *Einstein* x-ray telescope could have detected an x-ray burst equivalent to a dental x-ray, and the best infrared telescopes could detect the infrared radiation from a rabbit.

These examples make two points. One is the sensitivity of the new telescopes. Astronomy is a science that pushes technology and imagination to the limit. This is the key to its lasting popular appeal. The other point is that astronomy is no longer confined to visible light. Today, for the first time, the universe is being studied with almost equal sensitivity over a wide range of wavelengths, from the longest radio waves to the shortest gamma rays. The result is a much richer perspective on stars, galaxies, and other forms of cosmic matter. To understand why, we might imagine what we could learn from a local radio station if visual light were the only radiation we could detect. We would see a tower of steel and aluminum, but we would never suspect that it was broadcasting the daily news, weather, stock reports, or music as diverse as Pachelbel and Willie Nelson. Add a radio receiver to our capabilities, and suddenly the world around a lonely radio tower is a much more interesting place to be.

Visible light, or optical radiation, is part of a large family called *electromagnetic radiation.* This radiation is usually produced when electrons, the tiny charges on the outer parts of atoms, undergo rapid changes in their state of motion, or energy. These rapid changes, which can be thought of as vibrations, produce bundles of energy called photons. The photons move away from accelerating electrons at the speed of light. (All electromagnetic radiations move at the same speed in a vacuum.) In general, the more rapid the vibration, the more energetic the photons. For example, the vibration required to produce an x-ray photon is several hundred to a thousand times more rapid than that needed to produce an optical photon; in turn, a vibration millions of times more rapid is required to produce an optical photon than a radio photon.

Streams of photons behave in many ways like the water waves in a bathtub. The more rapidly your hand moves back and forth, the higher the frequency of the waves. Like water waves, electromagnetic waves have frequencies that depend on how rapidly the electrons producing the waves accelerate. This means that a high-frequency wave is composed of high-energy photons, a low-frequency wave of low-energy photons. The number of photons contained in a wave is not necessarily related to the frequency of the wave; it depends on the amplitude or intensity of the wave (see table, p. 210).

A characteristic of waves closely related to frequency is wavelength—the distance between crests of the wave. If all waves travel at the same speed, then a wave with a wavelength of 10 centimeters will take twice as long to pass an observation point as a wave with a wavelength of 5 centimeters. As the wavelength decreases, the frequency increases proportionately. Thus, x-rays, which have high frequencies, have short wavelengths; and radio waves, which have low frequencies, have long wavelengths.

Electromagnetic waves come in all wavelengths, but our eyes are sensitive to only a very small portion of the total spectrum. On the long-wavelength (low-frequency) side of the optical portion of the spectrum we encounter first infrared, then microwave, and finally radio radiations as wavelength increases. To the short-wavelength side of visible light lie the ultraviolet, x-ray, and then gamma radiations. It is certainly no accident that our eyes are sensitive only to optical radiation. For one thing, the sun—the strongest source of electromagnetic radiation in our cosmic neighborhood—emits most of its radiation in the optical range, and so eyes sensitive to these wavelengths evolved in humans and a variety of other animals. Furthermore, water vapor, dust, ozone, and other molecules in our atmosphere absorb most of the other wavelengths of radiation.

This atmospheric screen does not absorb radio frequencies, however, and it is there that the exploration of the invisible universe began in earnest in the late 1940s and early 1950s. Using war-surplus radar dishes, radio astronomers soon discovered, to nearly everyone's surprise, that exploding stars, called supernovae, were strong sources of radio waves. A detailed study of these sources showed that the radio emission could be explained only if supernovae produced tremendous amounts of energy in the form of high-energy particles. Then followed the discovery that entire galaxies consisting of billions of stars are wracked from time to time with awesome explosive activity that can rearrange the entire galaxy and affect the space around it over distances as great as a million light years. In the 1960s and early 1970s detectors placed above the earth's atmospheric blanket aboard rockets and satellites gave us a first glimpse of the x-ray universe. The discovery of multimillion-degree gas around exploded stars, neutron stars, black holes, and supergiant galaxies reinforced the view that violent events and high-energy processes play a crucial and quite possibly a decisive role in the structure and evolution of our universe.

The discoveries of radio and x-ray astronomy changed our conceptions of the universe. They also changed astronomy. Many of the old school of optical astronomers found themselves being pushed aside by a brash band of interlopers: radiophysicists trained during the war who were aggressively applying their talents to develop the radio channel for exploration of the universe, particle physicists who were entering x-ray astronomy, and theoretical physicists who were intrigued by the possibility that new physical laws might be needed to explain the cosmic violence. Giant radio dishes sprouted at Jodrell

Bank in England, at Fleurs and Molonglo in Australia, at Arecibo in Puerto Rico, at Owens Valley, California, at Green Bank, West Virginia, and at other locations around the world.

These radio observatories were complex, multimillion-dollar projects. The astronomer who would oversee their construction and operation would have to be more than an adept scientist and academician. He must also be a manager and an entrepeneur. He must direct the work of dozens of scientists, engineers, and technicians, he must sell his ideas to funding agencies, and he must defend his project against critics who seemed to be everywhere: in rival scientific camps, in funding agencies, and on various budgetary watchdog committees.

One of the pioneer radio astronomers, Sir Bernard Lovell, who conceived and developed the Jodrell Bank Observatory, had to weather a parliamentary investigation into the ever-increasing cost overruns. At one point Lovell confided to a friend, only half jokingly, that he might spend his retirement in prison as a result of the whole affair. Ironically, Lovell's faith in the importance of a large radio observatory was vindicated by an unexpected turn of events that had nothing to do with radio astronomy. On October 4, 1957, Russian space scientists astounded the world by claiming to have put the first artificial satellite in orbit around the earth. Lovell turned the Jodrell Bank telescope toward the predicted position of *Sputnik* and quickly picked up the satellite's radio transmissions, verifying the Russians' claim. British politicians and the public were proud that Great Britain played at least a minor role in this historic event, and accepted thereafter the idea that large radio telescopes are a good thing.

This book is about Lovell's successors. In a series of interviews, we have asked some prominent astronomers how they accomplished their goals of building and operating large observatories, and why. Often what they did not say was as revealing as what they said. Astronomy today is a tightly knit community in which news travels rapidly; a mistaken impression here, a bruised feeling there can change a key vote on a committee and jeopardize a consensus that a scientist-politician has been working years to build. In spite of this, many of those interviewed were quite frank and their comments have helped us to construct a montage of the tribulations and triumphs of the dreamers, the pragmatists, and the politicians who have been the driving forces behind the creation of the preeminent telescopes of the 1980s.

_ONE

The Mushrooms of
San Augustin:
The Very Large Array

_1

When we asked who was the driving force behind the development of the Very Large Array, the most powerful radio observatory in the world, we got an unequivocal answer.

"David Heeschen," said Barry Clark, a radio astronomer who has been working on the Very Large Array for twenty years. Clark is a native of west Texas and, like many of his fellow west Texans, a man of few words. But on the subject of David Heeschen and the VLA, Clark was not taciturn. "He was very good at keeping [the idea] going. He kept finding money during the dry spells, and kept interest high among the scientists. He is the person the community owes this machine to."

When we called to arrange an interview with this driving force, this man who sustained and guided the development of the best radio astronomy observatory in the world for sixteen years, we were surprised when he answered the phone. No protective secretary to screen calls, to inform us that he was in conference, to take our number for a return call. None of that. Just soft-spoken David Heeschen on the phone, agreeing to meet and talk with us about the VLA two weeks hence in Charlottesville, Virginia. Charlottesville is the headquarters of the National Radio Astronomy Observatory, the parent organization that administers the Very Large Array as well as several other radio telescopes around the United States.

Heeschen had resigned from the directorship of the National Radio Astronomy Observatory in 1978 after sixteen years, but ex-directors tend to keep many of the trappings of the office. Not David Heeschen. When we arrived in Charlottesville we found him, not in a spacious corner office, but in a small room midway down the hall. No sofa, a view that could only generously be described as pleasant, and no evidence—framed parchment, gilded ashtray, scale model—to suggest that he had ever been the director of the National Radio Astronomy Observatory and the father of the Very Large Array. Heeschen's style seemed to be too spare for such things. The shelves in his tidy office were not overflowing with technical books and journals. One was devoted to this year's copies of the *Astrophysical Journal,* another to green looseleaf binders, another to magnetic tapes, a box of Kleenex, and a model car—an Oldsmobile turbodiesel that has no particular significance, although he does like cars and used to race sports cars long ago. Other shelves contained neatly stacked manila folders which took up only a fraction of the available space. His desk was half covered with a computer terminal but was otherwise uncluttered.

We introduced ourselves to this tall, slender, casually dressed middle-aged man with sandy hair and pale blue eyes. After introductions, we asked how he came to be a radio astronomer.

"It happened more or less by accident," he replied. "I came out of the Army in 1945. I knew I wanted to go to college [and] I had already decided to go to the University of Illinois. I was from that region and my father went to school there, so no other school ever occurred to me. But I didn't know what I wanted to study. I didn't know that I wanted to be a scientist. I don't think I had an overwhelming interest in science."

"My father was a chemist. I enjoyed the science I had in high school, so I guess I had some kind of inkling, but I certainly didn't have a firm conviction. In fact I was sort of interested in agriculture. But the agriculture curriculum in the university turned me off. I was good in math, so I turned to science, because it looked more interesting. The most exciting-looking program was engineering physics, so I took it. Along toward the end of my undergraduate program I took some astronomy courses—simple-minded descriptive courses—and I liked them. So I hung around the University of Illinois for a couple of years after I got my degree, pursuing astronomy." When Heeschen finally decided to get serious about astronomy, he applied and was accepted to Harvard University's graduate program.

Around the time Heeschen arrived in Cambridge, a new field of astronomy was developing, and Harvard College Observatory, for the first time in years, was at the forefront. New England, with its humid climate, cloudy skies, and high population density, had never been the ideal place for a world-class telescope, but the concentration of intellectual talent in Cambridge had kept Harvard College Observatory in the vanguard of astronomical research throughout the nineteenth century. After all, contrary to popular notions, astronomers do not spend all their waking hours peering through telescopes. They only have to use them several weeks, or in some cases several days, a year. In fact, they need not visit a telescope at all; a Harvard astronomer can sit in his office on Garden Street not far from the mansion of Henry Wadsworth Longfellow and analyze the results of photographic plates taken by astronomers in the field.

Edward Charles Pickering, who became director of the Harvard Observatory in 1877, appreciated this. He set up observatories in Flagstaff, Arizona, and Arequipa, Peru, where the air is clear. Astronomers at these "stations," as they were called, would photograph the northern and southern skies and send the exposed plates to Cambridge. Eventually a definitive catalog of the spectra of stars, the *Henry Draper Catalog*, was compiled. Henry Draper, an accomplished amateur astronomer as well as a physician, did none of the work on the catalog; the title acknowledges the endowment from his widow that funded the research. The project was directed by Pickering, but most of the actual work was done by the legendary Harvard women, two score or more of underpaid and largely unrecognized "women computers," whose patient analysis and classification of photographic plates laid the foundation for many of the exciting developments of early twentieth-century astronomy. For example, the work of Antonia Maury and Annie J. Cannon led to an understanding that stars evolve through a sequence of well-defined phases in the course of their existence, and the work of Henrietta Leavitt allowed Harlow Shapley to make the first realistic estimates of the size and structure of our galaxy.

Harvard astronomers did not have to go to the mountaintop to do first-rate astronomy; the mountain could be brought to Harvard. Nevertheless, many astronomers like to be near the big telescopes, partly so they can gain more frequent access to them and partly because they have more to say, through positions on advisory committees, about how others use them. For these reasons, when Harlow Shapley was offered the position of director of Harvard College Observatory in 1920, friends urged him to reject the offer and keep his position at Mt. Wilson Observatory, at that time the home of the world's best telescope. The great astronomical discoveries would be made at Mt. Wilson, not at Harvard, his friends argued. Shapley chose Harvard, and for a person with his broad spectrum of talents and ambitions, he doubtless made the right choice. As president of several scientific societies, as well as a widely read popularizer of science, he became an influential spokesman in government circles for the cause of science and was instrumental in the founding of

UNESCO. At the same time, he continued to be a prolific scientific researcher and to attract some of the best astronomers in the world to Harvard.

But if Shapley had been right about his own particular career, so had his friends at Mt. Wilson been right about the future of astronomy. As Edwin Hubble and others explored the realm of the galaxies with the 100-inch reflector on Mt. Wilson, and showed for the first time that we live in an expanding universe, the center of astronomical research shifted west to Pasadena, the home of the California Institute of Technology, where the Mt. Wilson Observatory is headquartered. The addition of the Mt. Palomar Observatory, with its 200-inch reflector, brought the best and brightest of the optical astronomers to Cal Tech for the next three decades.

Shapley's thirty-year reign at Harvard was coming to an end when Heeschen arrived in 1951. And so was the absolute reign of optical telescopes as a means for studying the universe. In the 1930s and early 1940s, American radio engineers Karl Jansky and Grote Reber had pioneered a new field of astronomy. They had demonstrated with relatively crude receivers that the center of our galaxy and certain other inconspicuous optical objects were strong sources of radio waves.

After the Second World War, the Europeans, especially the English, and the Australians entered the field of radio astronomy and quickly transformed it into a major new area of research. In 1946 J. S. Hey and his colleagues in England detected radio waves from a source that was later identified with a distant galaxy. In 1948 another pair of English radio astronomers, Martin Ryle and Graham Smith, detected a more powerful radio source that was later identified with the remnants of an exploded star, or supernova. Shortly thereafter the Australian radio astronomer John Bolton and his colleagues detected four more sources of cosmic radio waves; these were later identified with supernova remnants and violently active galaxies. In the early 1950s surveys of the sky by Ryle and his colleagues in England, by B. Y. Mills in Australia, and by John Kraus and his colleagues at Ohio State University detected scores of cosmic radio sources. More often than not, they were located in regions of space where violent, explosive activity was occurring. This activity, which had for the most part escaped detection by optical telescopes, was far more widespread than had hitherto been suspected. Clearly, radio astronomy would make important contributions to our understanding of this violent face of the universe.

But radio astronomy would also help to answer the more traditional questions of astronomy: What is the size and shape of our galaxy? How rapidly is it rotating? How many stars does it contain? In 1944 Henk van de Hulst, a young Dutch astronomer, presented a paper at the University of Leiden which showed that hydrogen atoms in interstellar space could be emitting radio waves as a result of transitions between two very closely spaced energy states of the electrons in a hydrogen atom. For a hydrogen atom at rest, this radiation would be emitted at a wavelength of 21 centimeters. If the hydrogen atoms were moving toward the earth, then the radiation would be shifted to slightly shorter wavelengths; if the hydrogen atoms were moving away from the earth, then the radiation would be shifted to slightly longer wavelengths. This is an example of the Doppler effect; a more familiar example is the rising, then falling, pitch of a siren on a vehicle that approaches and then recedes. Since hydrogen is by far the most abundant element in the universe (80 percent by mass of the total mass of the elements), van de Hulst reasoned that any gas that exists in our galaxy would be predominantly hydrogen and should produce detectable 21-centimeter radiation. Furthermore, by studying the detailed distribution of radiation according to wavelength, it should be possible to tell how much hydrogen gas one is looking through and how it is moving.

Van de Hulst's suggestion opened the window to the galaxy. If, as seemed feasible with only a small advance in technology, the hydrogen 21-centimeter line could be observed, then it would be possible to investigate the motion of clouds of interstellar gas, and hence the rotation of the galaxy. Such observations are much more difficult to perform at optical wavelengths, because interstellar space contains dust particles that absorb optical radiation from remote regions of the galaxy. Radio waves do not have this problem. In addition, radio observations can be carried out at any time of the day and in almost any weather, whereas optical observations require clear night skies.

Because of the Nazi occupation of Holland, news of van de Hulst's idea spread slowly, and it was not until after World War II that the international astronomical community began to appreciate its significance. Then a race began to detect the 21-centimeter line. The

Dutch were the first to enter, but their receiver was destroyed in a fire. Across the Atlantic, a Harvard graduate student in physics, Harold Ewen, told Professor Edward Purcell about the 21-centimeter line. Purcell, who had worked on radar during the war and had a reputation as one of the best experimental physicists in the world, said, "Let's see if we can do it." Spending $400 on materials, they constructed a small horn-shaped radio antenna about the size of a bathtub and mounted it on the roof of the physics laboratory. On March 25, 1951, they detected 21-centimeter radiation. Six weeks later, the Dutch team confirmed the Harvard work, and by July an Australian team had also detected 21-centimeter radiation from the galaxy. Radio astronomers had acquired a valuable new tool. The Australian and Dutch teams, with larger, more versatile antennas than the Harvard group, immediately began to make detailed 21-centimeter maps of the galaxy—which clearly showed its spiral structure and allowed a rough determination of both the total mass and the proportion of hydrogen gas in the galaxy.

Meanwhile, the group at Harvard did not press their obvious technical advantage in this hot new branch of astronomy. Purcell, who would win a Nobel Prize in physics the next year for work unrelated to his achievements in radio astronomy, was first of all a physicist; astronomy was more a sideline to be pursued when bright students came along. Ewen, who had been one of Purcell's bright students, seemed more interested in the commercial possibilities of the new technology he had helped to develop, and had formed a company to produce radio receivers. Bart Bok, a Harvard astronomer and native of Holland, had followed developments in radio astronomy with interest, in part because some of the best work was being done at his alma mater, the University of Leiden, but primarily because of the importance of radio astronomy for understanding the structure of the galaxy, a subject to which he had made fundamental contributions as an optical astronomer.

It took, finally, a graduate student to push the Harvard Observatory into radio astronomy. Campbell Wade, who had left Harvard to join the army during World War II, returned to pursue his degree shortly after the 21-centimeter discovery. He went to Ewen and asked if he could work with his radio equipment.

"Ewen was a bit of an opportunist," Heeschen recalled. "He said, 'Fine, but I'm not going to spend my time teaching one student. Get someone else interested and I will do it.' So Wade went to Bok, who went to Ed Lilley and me. Because I had a background in physics and knew some electronics, Bok got the impression that I was good at electronics. That's how I got into radio astronomy."

The three students returned to Ewen, ready to learn the basic techniques of radio astronomy. Ewen then said that if Harvard was going to get serious about radio astronomy, they had to get a radio telescope. Under Bok's tutelage, Heeschen, Lilley, and Wade did the legwork for the project, and Harvard Observatory soon had a 24-foot radio telescope, which Ewen taught the graduate students to use. "Then," Heeschen recalled, in a tone that, twenty years after the fact, still reflected amazement at a mind so obviously unlike his own, "he informed us that he had formed a company and sold the receiver to NRL [the Naval Research Laboratory in Washington, D.C.]. So . . . we had a radio telescope without a receiver." Ewen, of course, had a solution. "He told us he would be happy to help us build a new receiver," Heeschen said. "So his company sold yet another receiver to Harvard."

Once they had a radio telescope, Heeschen, Lilley, and Wade set to work on independent projects. Over the next few years they each made major research contributions and they helped to build a 60-foot radio dish for Harvard Observatory. This radio telescope is located in Harvard, Massachusetts, a small town about 30 miles west of Cambridge. In the early fifties it was one of the major radio telescopes in the world; with it Heeschen, Lilley, Wade, and others established themselves and Harvard as stars in the firmament of radio astronomy. Although it has now been superseded by larger, more sophisticated telescopes, the 60-foot telescope is still operational; through a grant from the Planetary Society, it is used to search nearby stars for radio signals from extraterrestrial intelligences.

It was apparent that Harvard's eminence in the field of radio astronomy would be short-lived unless they built even larger telescopes that could compete with those under construction in England, Holland, and Australia. But a competitive radio telescope would be prohibitively expensive—too expensive for a single institution. Bok and other astronomers began to campaign for a national observatory dedicated to radio astronomy. The observatory would be supported by a group of universities and the facilities would be available to scientists from any institution. The newly formed National Science

Foundation endorsed the idea and contracted with Associated Universities, Inc., to build and operate the National Radio Astronomy Observatory. Associated Universities, Inc., a nonprofit corporation sponsored by Columbia, Cornell, Harvard, Johns Hopkins, Massachusetts Institute of Technology, Pennsylvania, Princeton, Rochester, and Yale universities, needed a young radio astronomer to help them with the planning. Acting on Bok's suggestion, they chose David Heeschen, who became the first scientist to work for the new observatory. By 1962 he had succeeded Otto Struve as director of the National Radio Astronomy Observatory.

2

An optical telescope requires clear air and dark nights. The growth of Los Angeles made it impossible to observe extremely distant and faint objects from the Mt. Wilson Observatory, and the growth of San Diego threatens the usefulness of the Mt. Palomar Observatory. In a similar way, high-precision measurements of faint radio signals from distant galaxies require an environment with very low levels of man-made radio noise. Aircraft, automobiles, power saws, televisions, and other household appliances can cause problems. The worst problems come from radio and television transmitters. Signals from a relatively low-power local radio station completely overwhelm cosmic radio waves. Therefore, radio astronomers must avoid the wavelengths at which radio and television stations broadcast; indeed, earthbound radio astronomy could not exist if certain wavelength bands were not off-limits to radio and television stations. About two dozen bands have been allotted to radio astronomy by international convention, but conventions change and radio astronomers must continually battle against proposed encroachments onto their electromagnetic territory.

When the first National Radio Astronomy Observatory telescopes were being planned, a primary concern was to find a site where man-made radio noise was low. They found it in Deer Creek Valley near Green Bank, West Virginia. Deer Creek Valley is protected by its isolation and by the Allegheny Mountains from the electromagnetic hisses, crackles, and rumbles of industry and commerce. This protection was secured when the Federal Communications Commission agreed to set up a radio-quiet zone around Green Bank, and the State of West Virginia passed an act designed to keep locally generated radio noise to a minimum.

Under Heeschen's leadership, the National Radio Astronomy Observatory became the focal point for radio astronomy in the United States. An 85-foot radio telescope that had been installed in 1959 was supplemented by a 300-foot telescope in 1962. But the technology of radio astronomy was changing as rapidly as the pace of new discoveries. In 1963 Cornell University radio astronomers and engineers completed a mammoth 1,000-foot radio telescope at Arecibo in Puerto Rico. By building the telescope into a large limestone hollow, the Cornell team had cleverly circumvented the problem of a costly support structure. But for all its grandeur, the Arecibo telescope has a serious drawback. It can scan the sky as the earth rotates, but it cannot be steered. The Green Bank telescope could be tipped up and down, but it too had to rely on the rotation of the earth to move it east and west. A fully steerable 140-foot telescope for Green Bank was in the development stage, but Lovell's 250-foot masterpiece at Jodrell Bank in England appeared to be near the limit of practicality. A similar limit had stopped the development of optical telescopes. It simply was not practical from a technical standpoint to build a reflecting mirror much larger than 200 inches in diameter; so the development of optical telescopes essentially stopped with the Mt. Palomar mirror. Had radio astronomers come up against a technological barrier that would not be overcome for decades?

Martin Ryle and his colleagues at Cambridge University in England had already given the answer. It may be practically impossible to build an enormous monolithic radio telescope, but a large—very large—radio telescope can be synthesized from several smaller ones. By linking the smaller telescopes electronically, their signals can be combined to produce a map of a cosmic radio source such as a galaxy that is far superior to the maps that can be made with any one of the individual telescopes. But the map is still not perfect because the synthetic telescope is full of blank spaces, caused by the spaces between the smaller telescopes. Ryle's ingenious solution to this problem was to make one or more of the smaller telescopes movable. By varying the spacing between the telescopes, and adding all the signals together, the "aperture" of the synthetic telescope could be filled in, and the performance of the synthetic telescope could approach that of a single, monolithic telescope.

By 1958 Ryle and his group had used these principles to construct one of the best radio astronomy observatories in the world. The One Mile Radio Telescope at the Mullard Radio Astronomy Laboratory, outside of Cambridge, England, consisted of three steerable 60-foot dishes placed along an east–west railroad track one mile long. Two of the dishes are fixed at either end of the track, while the third is mounted on a diesel tractor that moves it back and forth along the track. By recording the signals at each of 128 spacings, an aperture corresponding to a telescope one mile in diameter is synthesized. The use of three telescopes made it possible to make observations at two spacings simultaneously. Of course three telescopes in a line make for a long, narrow synthetic telescope. But as the earth rotates, the string of telescopes swings through 360 degrees in space, so that all angles are covered. In half a day, an oval-shaped ring can be synthesized. To fill in the middle of the ring requires observations at all different spacings, or 64 days.

Ryle's aperture-synthesis technique was quickly perceived as the wave of the future in radio astronomy, and radio astronomers around the world quickly made plans for their own "supersynthesis" telescopes, as they came to be called. The Dutch built an array of fourteen dishes along 3 kilometers of railroad track at Westerbork near Groningen in 1970, and Ryle directed the construction of the Five Kilometer Telescope with eight dishes at Cambridge, which was completed in 1972. As early as 1961, Heeschen and others at NRAO began making plans to get into the aperture-synthesis game.

Although radio astronomy was still in its infancy in the early sixties, it clearly had much to contribute to virtually every field of astronomy. The explosions on the surface of sun known as solar flares give off intense radio bursts containing vital clues to the nature of the underlying processes. The remnants of exploded stars were discovered with radio telescopes, and detailed studies showed that vast clouds of high-energy electrons were somehow generated in the explosions. In some cases, entire galaxies were found to be surrounded by such clouds; hundreds of these "radio galaxies" were discovered in a comprehensive survey made at Ryle's Cambridge radio astronomy observatory. Also found in this survey were the enigmatic "quasi-stellar radio sources," or "quasars," as they came to be known. Today there is a general acceptance among astrophysicists that quasars are the extremely energetic nuclei of galaxies, powered, many believe, by the infall of matter into a supermassive black hole. (A black hole is a region in space, formed by the collapse of a massive star, where the gravitational force is so strong that once matter has fallen inside, it cannot escape.) In the early sixties no one had more than the vaguest idea what quasars were. Some believed that an overhaul of existing theories of physics would be required to explain them. To a radio astronomer, the challenge was clear: find more of them, locate them more accurately so that identifications with optical objects could be made, and find out if they had any extended structure or were really star-like objects.

The accelerating pace of discoveries such as these convinced Heeschen that he and his colleagues at the National Radio Astronomy Observatory (NRAO) should not err on the side of modesty or caution in their planning. "The NRAO design group tended," he recalled, "to take an ambitious attitude toward the proposed instrument . . . relatively little was known at that time about radio sources, galactic or extragalactic, but it was clear that they embraced a very wide range of properties and characteristics. Any particular existing radio telescope, on the other hand, could measure only a small range of properties . . . It seemed inappropriate to restrict the performance goals for the contemplated instrument until we were forced to do so for technical or economic reasons."

They wanted a radio telescope that could make maps or images of objects that would compare favorably with those made with large optical telescopes. This goal immediately put their proposed telescope in a class by itself. It would have to have a resolution 20 times better than the Cambridge One Mile Telescope. This meant that it would have to spread across more than 20 miles. It also meant that a single large telescope was out of the question. The only way to go was to use the aperture-synthesis technique, with its array of movable telescopes. But how many telescopes would be needed in this "very large array," or VLA, as they began calling it? What size should they be? In what configuration should they be arranged? To help answer these and other questions about the design of the VLA, Heeschen acquired funds from the National Science Foundation in 1964 to build a second 85-foot antenna at Green Bank that could be moved along a railroad track.

One factor that limits the performance of any earth-bound telescope is the atmosphere. Atmospheric irregularities along the light path cause different rays to take slightly different paths to us. This is caused by the refraction, or bending, of light rays as they pass through small atmospheric "clouds." For visible light, these irregularities show up in the twinkling of stars, for example. Such fluctuations tend to average out for sources that are close enough to earth to appear large; the moon or even the planets do not appear to twinkle. But for sources with a small apparent diameter, twinkling is a frustrating problem over which the astronomer has little control short of putting the telescope above the atmosphere—in space. In a similar way, irregularities in the earth's upper atmosphere produce a twinkling of radio sources. In the mid-1960s little was known about the magnitude of such effects (called atmospheric phase fluctuations) and whether or not they would get much worse for a large array of antennas.

The Green Bank interferometer was used to study this question. A third 85-foot antenna, which could also be moved along the railroad track, and another, smaller, portable antenna were added. The portable antenna was carried to a distance of 35 kilometers, to test the feasibility of the VLA concept. The Green Bank interferometer was of fundamental importance in all aspects of VLA development. It gave scientists and engineers direct experience in the design, operation, and use of much of the equipment and many of the techniques that would be used in the VLA. For example, to test a possible alternative to moving the antennas on railroad tracks, the scientists tried placing the two movable antennas of the Green Bank interferometer on rubber-tired wheels, which allowed them to be pulled from one location to another. This scheme met with a number of difficulties. The road bed was expensive to maintain, the cost of the tires was high, and the heavy antennas were difficult to pull without tipping them over. It soon became clear that steel wheels on railroad tracks were unquestionably more reliable than rubber tires on a roadbed. The interferometer was also used to test many of the electronic systems that were ultimately used in, or rejected for, the VLA. Techniques for the analysis of data were developed and tested.

Apart from its usefulness in designing the VLA, the Green Bank interferometer was a powerful scientific instrument in its own right. It made highly detailed maps of radio sources and very precise measurements of their positions. Since the inauguration of the VLA, the Green Bank interferometer has been used exclusively to measure the relative positions of quasars and other small radio sources. This program is carried out by the NRAO staff for the United States Naval Observatory for the purpose of monitoring the rotation of the earth and the movement of the north pole. This is done by measuring the precise time at which each radio source passes overhead. If the earth were rotating at the same constant speed around a fixed axis, the observations would give the same time of passage for the sources each day. However, because of winds, tides, earthquakes, the movements of the earth's plates, changes deep in the interior of the earth, and the gravitational effects of other planets, the earth is not rotating at a constant rate, and its poles are wobbling slowly. This information is valuable for geophysicists; it is also of immense practical value. Astronomers need to know the magnitude of these effects in order to point telescopes accurately, as do navigators who use the stars and artificial satellites to determine their positions at sea. Space scientists use this information to guide space probes through the solar system.

By 1967, after three years' experimentation with the Green Bank interferometer, the design group had agreed on many of the details of the VLA concept. It would consist of an array of 27 antennas, each 25 meters in diameter. Computer simulations by Leonard Chow and others had shown that improvement in the quality of radio images increased only slowly for more than 27 antennas, whereas the cost climbed proportionately to the number of antennas. The antennas would be positioned in a Y-shaped configuration, with each arm of the Y about 21 kilometers in length. The Y configuration would make it possible to synthesize maps in a much shorter time than the 12 hours required for linear arrays. Every antenna would be individually connected to every other antenna, giving 351 separate pairs.

We asked Heeschen how they decided to put the VLA on the Plains of San Augustin in New Mexico. "It [the selection of the site] was actually pretty straightforward, because the requirements for the site are fairly stringent. You needed an area at least 35 kilometers in diameter. It had to be relatively flat so you could move the antennas around easily. That much flat ground that is available and not being used is hard to come by." In fact, the Plains of San Augustin had attracted the attention of other radio astronomers. "Bill Erickson at Maryland was interested in low-frequency arrays and had his eyes on it two or three years before we did," Heeschen said.

Did they find the site just by looking on maps? "Yeah, that's the way we found it," Heeschen said. "A search was made with topographical maps for anything that was large enough and didn't have a mountain in the way. When those were found, and there were only forty or forty-five of them west of the Mississippi [this was a requirement designed to keep the VLA remote from large population and industrial centers], then people began to go out and visit them. Some of them were full of oil wells, some were old lava beds, some were heavily populated, some were very, very isolated. We originally confined the search to a region south of forty degrees latitude, because the array works better at low latitudes. Best on the equator, in terms of sky coverage and other parameters. So we wanted to go as far south as we could."

He smiled as he remembered an exception. "When we couldn't get money, we heard about the Fleischmann Foundation. Rumor had it that they had a lot of money, about a hundred million dollars, and that they had to use it. It seemed to fit beautifully into the needs of the VLA. So we decided to visit them. Their headquarters is in Nevada. We thought we should be prepared to find a site near there. We expanded the search and found some beautiful sites in Nevada and Utah. Too far north, though. And extremely isolated. It turned out that the Fleischmann Foundation wasn't interested, and if they had been, they wouldn't have cared where the site was, so we went through that little effort for nothing. Anyway, of the forty sites, just looking at them eliminated all but seven. Those seven were studied in considerable detail. This one site [the Plains of San Augustin] is just head and shoulders above the others. It is at higher elevation, has easy access. Only five or six landowners. It is almost ideally suited for a large radio array."

Certainly, from purely aesthetic considerations, the site is ideal. A few miles outside Magdalena, an old cowtown in central New Mexico, this ancient lakebed had once been the home of Folsom man, who gathered on the banks of the lake to fashion spear points. When we drove out to see the VLA, we stopped on a hill several miles away, to get an overall view of the array—over two dozen huge, gleaming-white radio dishes strung out like giant mushrooms on the desert floor. Though it was early September and was uncomfortably warm in Soccorro and Albuquerque, we shivered in the cool, high desert wind. Probably we would have shivered anyway. Observatories do that to us, for much the same reason that our first views of Stonehenge, the Parthenon, and a Van Gogh painting raised goosebumps. They represent the reaching beyond ourselves—not for power or wealth, but for the delight of it—that gives the human species its moments of grandeur. As we stood beside U.S. Highway 60 and looked at the Very Large Array of radio telescopes on the Plains of San Augustin, it was impossible not to sense that grandeur and shiver.

3

Despite the diligence of the planning and the perfection of the site, there were no guarantees that the Very Large Array would ever exist except as a dream. Other groups of radio astronomers had dreams, too. A group at the California Institute of Technology had plans for an array of eight antennas in Owens Valley in central California. A consortium of scientists from thirteen New England universities and observatories, including Harvard University and the Massachusetts Institute of Technology, had proposed a 440-foot steerable telescope built inside a "radome" to protect it from the wind and weather, and a Cornell University group proposed to put a huge, fixed radio dish in the ground in what was essentially an improved version of the Arecibo telescope. There was not enough money to convert everyone's dreams into reality, and the protracted battle for funds that lasted through the 1960s was bruising.

"There was a tremendous amount of fighting as to which project was going to get funding," Heeschen remembered. "The radio astronomers had all their ducks in order. Everybody had a telescope designed and had it well justified. All these possible telescopes were sitting there, waiting to be funded—all with years of past justification behind them and groups ready to go and so on. The fighting to get one of them funded was pretty vigorous."

When we asked him to elaborate, Heeschen made it clear that he had no desire to re-enact old battles. "I don't see any point in dragging that up. Some of the people are still around. It wasn't vicious fighting, because actually everybody played pretty fair. Once the decision was made, all the radio astronomers supported it."

The decision was a long time in coming. The first VLA proposal was submitted to the National Science Foundation in 1967. A committee, headed by Robert Dicke of Princeton University, was set up to evaluate the proposals and make recommendations to the National Science Foundation. The results of the Dicke committee's deliberations were not always to Heeschen's liking. "There were periods when we thought that some other group had gotten a much stronger recommendation out of the Dicke committee than we had." The NRAO design group continued to refine and improve the VLA concept for the next two years, and in 1969 submitted a new proposal, incorporating what they had learned. Still, the Dicke committee refused to come out strongly in favor of any proposal. Instead, it recommended that both the Owens Valley and radome projects be built at once and that the VLA be built in stages over a period of several years. Such a blanket recommendation was clearly impractical. So it was of little or no use to the supporters of the various projects. It was a low point for Heeschen.

"We had no idea when we were going to get funded, if at all. We cut the project off and disbanded the design group, because it was clear that whatever got built should be relevant to the time it got built, and we couldn't design it because there was going to be a long hiatus. We couldn't really design the hardware too far in advance. So we stopped. We had a good conceptual design. We had a hardware design that we knew worked. We were ready to go to working drawings in everything, but we stopped at that point, for about three years."

During this intermission, the fate of the VLA and the other proposals was decided not by the Dicke committee but by a committee of 23 eminent astronomers headed by Jesse Greenstein of Cal Tech. The Greenstein committee was established by the National Academy of Sciences to survey astronomy and astrophysics, determine what new instruments were needed, and establish priorities for funding in the 1970s. Although he was not a member, William Wright of the National Science Foundation played a pivotal role in helping the radio astronomy subpanel of the Greenstein committee to reach its decision.

Heeschen was chairman of the subpanel, which included two members from MIT, one from Cal Tech, and five other experts in radio astronomy. In its preliminary report, prepared in December 1969, the subpanel endorsed the recommendations of the second Dicke panel. Harvey Brooks, speaking on behalf of the National Academy of Sciences, urged the subpanel to establish priorities. Wright, meanwhile, let it be known that the National Science Foundation— the agency that would provide the funds—favored the VLA. Although the 440-foot telescope and the Owens Valley Array would be among the best radio telescopes in the world, the VLA would clearly be much better than the best. Pressured by both the National Academy of Sciences and the National Science Foundation to make a choice, the radio astronomy subpanel recommended the Very Large Array. Along with the recommendations of the subpanels, the Greenstein committee solicited the opinions of hundreds of astronomers, so that its report, *Astronomy and Astrophysics for the 1970's,* was widely viewed as a consensus among astronomers and astrophysicists.

"The radio astronomy subpanel made the VLA the number-one priority," Heeschen recalled with pride, "and the full committee made it the number-one priority for all of astronomy. The Greenstein committee recommendation was crucial. We had won the competition, both in the eyes of the radio astronomy community and in the eyes of the astronomy community as a whole."

Shortly after the Greenstein committee's recommendations were announced in the spring of 1971, Heeschen resurrected the VLA design group and appointed Hans Hvatum as chairman. In 1972 the National Science Foundation and Congress approved construction of the VLA. The National Science Foundation and the Office of Management and Budget agreed that the project should be funded at about $10 million a year. This stringent rationing of the rate of funding led to a longer construction period, and consequently to higher costs. It did have advantages, though.

"When we started again," Heeschen pointed out, "we could have taken the old design and made working drawings out of that. Instead, since three years had passed and there had been a lot of advances in electronics, there is little resemblance between the design that was proposed and the one that was used." The longer period for construction gave time for detailed redesign and testing. "So," Heeschen continued, "although we were unhappy at the time with the schedule imposed by the funding rate, in the long run it probably did no harm and perhaps did some good." At any rate, Heeschen emphasized, in an evocation of "heartland" principles that has probably more than once been mistaken for naiveté, "NRAO was determined to build the VLA on schedule and within budget."

To do this, Heeschen and his colleagues at NRAO would have to maintain tight control over the job, which would involve everything from building a railroad to setting up a complex electronic "brain" to run the VLA and handle the data. They would have to minimize design changes and extras in subcontracts and make sure that design specifications were fully developed in detail before construction contracts were awarded. They decided that NRAO would act as its own prime contractor for the VLA.

Heeschen's determination to stay within budget was soon tested by a chain of events beyond his control. The mideast war, the oil embargos, the sinking value of the dollar, and the acceleration of inflation forced cutbacks. The building program was drastically modified. Except for the control room and the cafeteria, prefabricated buildings would be used. A proposed airstrip was eliminated. The number of transporters—the diesel-powered units that would move the antennas along the track—was reduced from three to two, and plans to equip the VLA with the capability of operating simultaneously at two frequencies were dropped. These budget cuts were judged to have the least impact on the scientific usefulness of the array.

The announcement that the National Science Foundation was going to spend $76 million to construct the VLA on the Plains of San Augustin attracted the attention of a number of New Mexicans. The north arm of the Y was planned to extend 2 kilometers into one of the very few irrigated acreages in New Mexico outside the Rio Grande valley. The value of this land skyrocketed as soon as the owner learned of the VLA's need for it. How to deal with the greedy owner? "We simply cut 2 kilometers off the north arm, so that it stops just short of the irrigated land." The owner got nothing.

Then there were problems with cattle. "We were sued by one of the ranchers," Heeschen recalled with a chuckle, "on the grounds that putting the track down would disrupt their traditional paths to water and food, that they would not cross the railroad tracks." The resolution? "We have pictures of cattle crossing railroad tracks. He lost that suit."

They had to worry not only about the present and the future but the past as well. "New Mexico law requires that all proposed construction areas be surveyed for the existence of sites of special archaeological interest," Heeschen said. "If any sites are found, they have to be investigated before construction work begins. Well, they found a site along the southwest arm of the array, and it was determined that it should be excavated." But no one was willing to pay—not the State of New Mexico, nor the National Science Foundation archaeology program, nor any university in the state. "After a lengthy period of bickering, NRAO and radio astronomy had to finance a bit of archaeological research."

A team of archaeologists from New Mexico State University found thousands of spear points, bone fragments, pottery shards, and lithic tools, remnants of three cultures that had occupied the site. The oldest was the Folsom culture, of about 10,000 years ago, followed by the Cochise culture around 1,000 B.C., and the Mogollon culture about 400 B.C.

In 1973, just before construction was to begin, the price of used steel rails soared far beyond the budgeted price. Heeschen felt that they had to avoid this added expense or the entire program might have to be canceled. "We made a desperate search for other sources of rail. We found out that there was a lot of abandoned railroad track on military bases around the country. We went to the NSF and they had some of this track declared surplus. It was made available to us for the cost of taking it off the bases and shipping it to New Mexico." The money saved by using the surplus rail enabled the NRAO to keep the scientific integrity of the VLA intact and to keep the cost very close to budget despite soaring inflation. The total cost upon completion was $78 million, less than 3 percent over budget during a time when the consumer price index for commodities and services increased by more than 50 percent.

We asked Heeschen if there were ever any critical moments in the construction. "There was a time," he answered, "when the whole array performance was threatened by sinkage of the waveguide." Roughly speaking, a waveguide is a metallic pipe that "guides" electromagnetic waves. The waveguides used at the VLA have an inner diameter of 6 centimeters, and carry radio waves that have wavelengths between 10 and 20 centimeters. At these wavelengths, waveguides are much more efficient than coaxial cable transmission lines. The pipes are pressurized with dry nitrogen to reduce attenuation by oxygen absorption and to prevent internal corrosion. A plastic jacket on the outside protects them from external corrosion. Because thermal expansion can present problems, the waveguides are buried a minimum of one meter underground.

The waveguides are, in effect, the "nervous system" of the VLA. When an antenna receives a radio signal from space, this signal is amplified, modulated on a carrier frequency, and transmitted back to the central control building through the waveguides. There it is compared with a signal from each of the other antennas. In addition, the waveguides must carry the signals that monitor and control the antennas and the electronics needed to process the signals, as well as reference signals that enable the antennas to determine precisely when they received a particular signal. All these signals are crucial for the operation of the array; the waveguides must carry them over distances of several miles with a minimum of loss. The major losses in a waveguide occur when the waves hit the sides of the guide. "The waveguides have to be terribly straight," Barry Clark told us. "The first piece of waveguide we installed was about a kilometer and a half long. We did careful measurements on the losses. The losses started to go up."

"It was settling too fast," Heeschen recalled, the memory still making him uncomfortable. "The losses were going up at a rate that was greater than was allowable. There was a period of uncertainty. Would it perform up to standards ten years from now? The method of burying it had to be changed." They compacted the trench, put more sand in, surveyed it again to make sure that it was straight to better than one part in a thousand, and tested it again. The initial performance was better, and the settling had stopped. "This was one of several instances," Heeschen emphasized, "where a critical task was carried out entirely under the direct supervision of NRAO personnel. By doing the job ourselves, we got better performance and quality control for a given cost."

We found this to be a common theme among the astronomers we interviewed. Despite their usual preoccupation with things on a cosmic scale, the observers are by and large a very practical lot. They know what they want and they want to have as much control as possible over building it. It is no coincidence, they maintain, that in the two outstanding success stories of recent times, the VLA and the *Einstein* x-ray observatory, the astronomers had such control, whereas in the trouble-plagued Space Telescope they did not.

_4

Scientific observations with the Very Large Array began in a small way as soon as the first two antennas were constructed in 1976. By 1977 six antennas were available, and by 1980 the entire array became operational. The 82-foot antennas are anchored to pedestals; they can operate in 40-mile-per-hour winds and are designed to survive gusts up to 125 miles per hour. Carl Bignell, the deputy director of the VLA (Ronald Ekers is the present director), told us that the strongest winds recorded so far at the VLA have been 90 miles per hour.

Bignell showed us around the VLA control building and explained the basic operational procedure. "Every few months the antennas are moved into a different configuration. Four basic configurations are used. The A configuration is the most extended. In it the antennas are spread along the full length of the arms, so that the outermost antennas are 20 miles apart. This enables the VLA to map small, intense radio sources with a resolution ten times that of the world's best optical telescopes. Configuration D, the most compact, brings the antennas together on very short arms, only four tenths of a mile long. There, maps of larger sources can be made at lower resolution but with very high sensitivity. This allows radio astronomers to study faint, extended fields of radio emission around objects such as relatively nearby remnants of exploded stars, or very distant but immense radio galaxies. Configurations B and C are intermediate between A and D.

The 215-ton antennas are moved from one foundation to the next by 90-ton transporter vehicles that move along the railroad tracks at speeds less than 5 miles per hour. A diesel electric generator mounted on the transporter provides power to critical components of the system during an antenna move. With two transporters, a change of configuration may require one and a half to five days, depending on which configurations are involved. Every seven weeks, one of the 27 antennas is taken out of the array for routine maintenance and repair, and a recently serviced antenna takes its place. In this way, each of the 28 antennas (27 plus 1 spare) is serviced every three and a half years.

The nerve center of the VLA is the control building, a two-story brownstone and concrete building of conventional institutional design. Except for the graceful, silvery Y-shaped sculpture "Shiva" by Jon Barlow Hudson that stands in front of the building, there is little to suggest the imagination of the people who work inside. Most scientific laboratories are that way. They have a proletarian, institutional drabness about them that resists the occasional efforts of their occupants to enliven their walls with posters or a scientific "joke" that perhaps two hundred people in the world would understand and less than a hundred would think funny. The laboratory spaces are usually more interesting, if for no other reason than they look vaguely familiar, like the garage or workshop of an enthusiastic but not especially neat hobbyist or ham radio operator. The workshop is on the ground floor of the VLA control center. There, in separate areas that seem to have grown by necessity rather than design, electrical engineers test and repair electronic components for the antennas and computers. It is there that the waveguides enter the building and swoop upward to the computer room. That day, the waveguides were bringing in signals that left a radio galaxy in the constellation of Lyra over a billion years ago. These signals would be fed into banks of computers upstairs.

In the control room on the second floor is a large instrument panel that resembles the air traffic control center for a small airport. Consoles are spread around a large, semicircular desk that commands a view of the VLA antennas. On the right-hand side of the desk is a computer that monitors over two hundred different critical items for each antenna. A flashing light summarizes the computer's assessment of the status of the VLA. Blue means everything is fine; yellow signifies minor difficulties; red is more serious; and magenta represents a four-star crisis. The light was flashing yellow. "A problem with the voltage regulator in one of the antennas," Bignell explained. "Nothing major, but we will have to keep an eye on it." If any one antenna goes out of operation temporarily, the performance of the VLA would not be optimal, but it would still be better than any other radio telescope in the world.

We asked what would happen in the event of a massive power failure—one that affected the entire central New Mexico area, for example. "Diesel engine generators are designed to come on within two minutes of a power failure," Bignell explained. "They stow the antennas, that is, they keep the critical components cool until the power failure is over. The same goes for any critical components in the control room."

To the left of the control panels is a large room containing the electronics that control the VLA. From here the waveguides carry, monitor, and control signals out to the antennas, and bring back the cosmic radio signals that the antennas have captured. These signals are amplified and digitized (converted to discrete numbers) and then passed through a delay line, so that the signals from all the antennas arrive at a common point in the proper phase. The time delays given to these signals are accurate to a few billionths of a second. Then the signals are split into 26 components and correlated with signals from each of the other 26 antennas. This correlation is the brain of the VLA's operation. It makes possible the synthesis of signals from 27 antennas that gives the VLA the performance characteristics of a single large telescope about 20 miles in diameter. For this to work, it is imperative that the correlators function properly. The room containing the correlators is encased in steel, so that no spurious radio waves, no electric fields of any kind, can leak into the room and jeopardize the performance of the correlators. Test signals are periodically put into the system and processed to ensure that the system is operating as it should.

The data from the correlators are transferred to an off-line computer which collects, sorts, and organizes them so that radio maps and other information can be extracted. Barry Clark directed the development of the computer system for the VLA. He freely admits that "the largest problem we have at the VLA is that we don't have enough computers to keep up with all the data processing we would like to do. It has turned out to be a much heavier load than we anticipated. There are some things we would like to do but just can't. The computer just bogs down, and it makes it slow and painful to do things."

Even so, Heeschen said, the data-processing system is "vastly improved over what it was. It is still the limiting factor. I wouldn't call it a major bottleneck anymore, but it is still the limiting factor in what you can do in certain areas, in spectroscopy, for example." Spectroscopy is the analysis of narrow peaks and valleys in the spectrum of a radio source—the 21-centimeter hydrogen line, for example. Observations in a large number of wavelength bands are generally required to get a good spectrum. Because of the computer limitations, scientists cannot yet use the full power of the VLA for this purpose. "They have to tailor their program," Heeschen said. "They can't observe a wide wavelength range, for example, or they have to use a limited number of antennas, which cuts down on the baseline coverage." But he was optimistic. "It's nothing like the problem it was. For a while it was a real problem. If you wanted to use the whole array, you could use only two or three channels instead of the whole frequency range. Or, if you wanted to use many channels, you could only use a couple of antennas."

We asked if the data-processing problem was the result of poor planning, or was it the lack of money that kept them from solving the problem ahead of time?

"The expectations of what people would do with the VLA increased tremendously between the time we first submitted the proposal and the time we eventually began building it," he explained. "There are a lot of different reasons for this. The science advanced, we changed the nature of the design to make it far more flexible."

One example of this is the use of the "snapshot mode." The instrument turned out to be so sensitive that high-quality maps of relatively strong sources can be made in 5 minutes, rather than the 8–10 hours required for most radio telescopes. So, instead of making one or maybe two maps in an 8-hour period, the observers are making hundreds of maps. This is a scientific bonanza, but it places a heavy burden on the data-processing system.

Another factor contributing to the increased load on the computer system was, Heeschen said, "a revolution in data processing. The techniques that are used today didn't exist at all in the sixties. The so-called 'CLEAN' techniques, the self-calibration techniques that exist today to take out some of the instrumental instabilities, hadn't even been dreamed of. These procedures are extremely valuable, but they increased the data-processing requirements by orders of magnitude. When we first designed the computer, we thought it was adequate for what we then thought the thing would do. But in the interim, so much happened that we simply couldn't afford."

This is not an uncommon occurrence in projects that depend on the very latest technology. The time between the proposing of a project such as the VLA and the funding of that project often stretches to several years. During this time, scientific knowledge expands, and important advances in technology occur. Or inflation simply drives up the price of goods and labor. The standard procedure is to ask for an increase in funding over the original request. Certainly in the case of the VLA it would have been well justified. But to ask for more money went against Heeschen's grain.

"We felt constrained to stick with the budget that we had sold the project for, and that any further changes ought to be compared with other things we might want to do with the money. So, we didn't make any attempt to add 10 percent of the cost to cover the computer capability." In the proposal, about 5 percent of the total cost was earmarked for computers and software. "The other reason is that I frankly don't see how we would have known what to do," Heeschen went on. "Things were evolving so fast and everyone was so uncertain as to the best way to handle it. The only thing we could've done is what I think NASA does, and that is a massive overkill, which would've been very, very expensive. And," he added, "not in our philosophy or our way of doing things at all. It would have been distasteful, I guess, besides being expensive. And I'm not sure we would've gotten such a good product. Now the thing has evolved in a natural way, along with peoples' understanding of what the instrument can do. Perhaps it has cost us a couple of years' time, but maybe the learning process has been useful in terms of designing what we really wanted. So I don't feel badly about it."

But other scientists did feel badly about it. They had a magnificent new radio telescope, yet it couldn't be used to its full capacity because of an inadequate computer. "There was a lot of unhappiness," Heeschen acknowledged, "because we had this instrument that could in principle make 200 maps at 200 frequencies with a high resolution, but the computer wouldn't handle the data. Yet," he laughed ironically, "I don't think anyone knew how to write the code for that, personally. And," he went on, "it wasn't a waste of time. The instrument was always fully utilized. The mix of jobs it could do changed as the capability to do spectroscopy increased. But when you couldn't do spectroscopy because of the computer, people doing other programs were happy because they had use of the instrument. It was still churning out good science, but not in the area of spectroscopy, so those people had to wait, that's all. They were very unhappy."

At the VLA, we asked Carl Bignell about the present status of the data-processing system. "As far as the VLA needs go, there's no question of our top priority. We need much better computer capability. In all areas, we have a big number-crunching problem and data-storage problem. We have to find a way of solving it."

Did he see it as a problem that will be solved in the near future? "I don't know," he replied. "There's a long-term computer plan, which is an attempt to say what we need in the way of computer capability. We will submit to NSF a special request, but it's not clear that we will get it. The VLBA [Very Long Baseline Array, a proposed array of ten 82-foot antennas spread across the United States from Alaska and Hawaii to the East Coast that are synchronized with atomic clocks] will have the same problem, so it may force the issue. The NRAO has always taken the approach that we can do a job for a certain amount of money by doing it the spartan way. It works, but it has its problems."

5

Despite the problems caused by its data-processing system, the large size and collecting area of the Very Large Array give it a unique combination of resolution and sensitivity. It is the world's most powerful directly-linked radio array and is likely to remain so for many years. Already, the VLA has been used to study virtually every known type of object in the universe. It has investigated the chemistry of the atmosphere of Jupiter, Saturn, Uranus, and Neptune. The VLA snapshot mode has made possible radio pictures of solar flares with a time resolution of 10 seconds and high spatial resolution. Solar flares release energy equivalent to a million hydrogen bombs. Observations of these events through the VLA snapshot mode have allowed solar physicists to pinpoint the source of microwave radiation released during a flare and to understand more about the shape of the magnetic field in the flaring region.

"I think the VLA has rejuvenated solar radio research," Heeschen said. "For years there didn't seem to be much exciting or interesting or new in solar radio astronomy. It seemed to have shriveled up. At least it seemed that way to me. There were very few people actually doing solar radio astronomy and those few often wished they would get more time. Partly at their behest, we tried to make sure that the VLA would be usable for the sun. It is very difficult, because the sun is so much brighter than other sources . . . the dynamic range that you force on your instrument by going from a very weak source to the sun is very tough. It causes all kinds of problems—for the receivers, for the calibration techniques, the software. It doesn't move at the same rate in the sky, for example—all these things cause problems. Well, they all got worked into the design. And the same goes for planetary research. They didn't get worked in as fast as some people would have liked, but a tremendous effort was made to make the VLA responsive to all possible areas where it could contribute. And I think the scheduling reflects that."

To get observing time on the VLA, prospective observers must write a proposal, including a scientific justification for what they want to do, and how much time they will need. This proposal is submitted on a standard form available from the NRAO headquarters in Charlottesville. From headquarters the proposal is sent to the

User's Committee at the VLA and to outside referees, who comment on it and rate it. Every three months, the User's Committee, which consists of three VLA personnel and the assistant director of NRAO headquarters, meets to decide which proposals will be given observing time in the next quarter.

We asked Heeschen if some proposals get more favorable treatment because they are more in line with the interests of the User's Committee members, or the overall philosophy of the director as to which lines of research are important.

"For years I was involved in this process and I think that most of our site managers [at Green Bank, the VLA, and the Millimeter Wave Radio Telescope on Kitt Peak outside Tucson, Arizona] try not to let their personal prejudices get too involved. Of course, it's inevitable that they will if you have a larger number of proposals than you can accommodate. If they are nearly equally rated by the referees, then these subjective matters are bound to enter. But we used to watch out for this. We tried to look at the statistics of what was happening from time to time and make sure that we were not getting too many prejudices involved. In particular, the users themselves, when they see the mix on the telescopes [a summary of the observation schedule for a given quarter is sent out to all prospective users], they will tell us whether they think the thing is being biased one way or another. Solar people, for example, wish they would get more time. At times they may have felt they were being discriminated against. We looked at it and usually decided that they were not. I don't think there is an undue emphasis on the part of Ekers [the present director] on extragalactic research. I hope not. I hope our people are more broad-minded than that. I believe they are. No system is perfect, and I'm sure that somebody else would wind up with a slightly different mix. I think that the mix also represents the degree of pressure on the telescope, the degree of interest in and the number of people involved in the various areas."

Barry Clark, who directs the scheduling at the VLA, agreed. "We are mostly driven by what the users want to do," he said. What they want to do is to use the VLA to study everything from Jupiter to the most distant objects in the universe. Some of the most exciting results have come from observations of regions where stars have recently formed, regions where stars have exploded, and the vicinity of what may be supermassive black holes.

Infrared and radio studies have provided us with essentially all we know about the formation of stars. With the VLA it is possible to study protostars, massive spheres of gas that separated from their parent clouds of gas and dust less than a few hundred thousand years ago. A protostar has an interior temperature sufficiently low (less than 10 million degrees) that nuclear fusion reactions cannot occur. Without a nuclear heat source in its core, a protostar slowly collapses under the inward pull of its own gravity. The energy released during this collapse will cause the interior temperature to rise until the nuclear reactions "turn on," and a star is born. The process by which a star forms is only poorly understood, partly because it occurs under cover, in clouds so dark and dusty that no optical radiation can escape.

Important questions remain unanswered. What causes the protostars to collapse out of their parent cloud in the first place? How do they solve the angular momentum problem? A figure skater's rate of rotation, that is, her angular momentum, rapidly speeds up as she pulls in her arms. In the same way, a star's rotational energy should rapidly increase as it collapses, to a point where it equals the gravitational energy causing the collapse and brings the collapse to a halt. Even very slowly rotating protostars should, in theory, have great difficulty forming. Yet apparently they do not. Several resolutions to this paradox have been put forward. Most of them involve the shedding of matter by the protostar into a ring or disk around its equator. This loss of matter would keep the rotational energy down. With the VLA, astronomers are compiling impressive evidence that rings or disks do indeed form around protostars. Was it from rings or disks such as this that the earth and other planets were formed when the protosun collapsed? Probably. If so, are planetary systems common around other stars? The VLA data, together with the results from the Infrared Astronomical Satellite (see Part IV, below), suggest that the raw material, namely the protoplanetary rings and disks, are common. One of the primary objectives of the Space Telescope will be to look for the finished product, namely planets, around nearby stars.

Some of the fundamental contributions of radio astronomy have had to do with the end, rather than the beginning, of a star's life. Through radio astronomy it was first realized that large quantities of extremely high-energy electrons with speeds near that of light are produced as a result of supernova explosions. The generation of extremely high-energy particles appears to be a universal phenomenon—not only in supernova remnants but on a far larger scale in quasars and radio galaxies, so an understanding of how energetic electrons are produced in supernova remnants has far-reaching implications.

The initial explosion cannot explain the high-energy electrons observed in supernova shells. Particles that did not quickly lose their energy as they cooled in the expanding cloud would have escaped from it long ago. Two supernova explosions are known to have left behind superdense neutron stars, which generate large quantities of high-energy electrons. At first it was thought that this would be the universal answer to the question of how high-energy particles are produced; it now appears that these objects are the exception rather than the rule. Many supernova remnants have no central neutron star; the acceleration of particles to high energies must have something to do with the shock wave produced by the supernova explosion.

X-ray images made by the *Einstein* observatory (see Part II, below) and radio images made by the VLA and other radio telescopes show that supernova shells tend to break up into many small fragments. It has been suggested that these fragments represent turbulent magnetic eddies, and that charged particles are accelerated to very high energies through collisions with the eddies. It is too early to say definitely that this process generates the high-energy electrons necessary to explain the radio emission in supernova remnants, and whether it will explain the radio emission from quasars and radio galaxies, but a preliminary analysis looks promising.

On the scale of galaxies, observations with the VLA are bringing astronomers ever closer to understanding one of the most perplexing riddles of modern astronomy—what is the energy source responsible for the violent activity observed in the centers, or nuclei, of galaxies? This activity is relatively modest in our own galaxy—out in the "galactic suburbs," where our solar system resides, some 30,000 light years from the center of the galaxy, we hardly notice it. But in quasars and other so-called "active galaxies," the violence has gotten out of hand. The appearance of the entire galaxy is affected, and in the extreme case of the quasars, the central source is so bright that it was almost twenty years before astronomers could detect the galaxy in which the quasar was embedded.

A bright quasar produces energy equal to that of a hundred trillion suns. X-ray observations indicate that this energy is produced in a compact region about the size of our solar system. Over the past decade, radio observations have been particularly important in establishing the nature of explosive activity in galactic nuclei and quasars. They have provided strong evidence that galactic nuclei and quasars explode repeatedly, ejecting clouds of high-energy particles, always along the same axis of symmetry. The radio jets extend out from the nucleus and point toward much larger bubbles or lobes of high-energy particles. In some cases these alignments stretch over tens and even hundreds of thousands of light years. These alignments imply that the ejection of high-energy particles has been channeled within a few percent of the same direction for the last several million years.

Such a well-defined axis of symmetry strongly suggests a single supermassive object as the source. But what type of supermassive object? The conventional theory of stellar structure, when applied to a star having the mass of a million or more suns, shows that it should collapse in a few thousand years to form a black hole. In normal stars, the gravitational forces are held in check by the intense heat generated by nuclear fusion reactions in the core of the star. When the central part of a star has used up its nuclear fuel, the core will collapse. If the star is about the same mass as the sun, it will turn into a white dwarf, a dense star about the size of the earth. A sample of white-dwarf material the size of a sugar cube would weigh ten tons. Matter at white-dwarf densities exhibits a property called electron degeneracy. The term degeneracy refers not to the moral character of the electrons but to the fact that they are crowded so closely together that all the low-energy states are filled, and some electrons are forced into higher-energy states. This manifests itself as a pressure, called electron degeneracy pressure, which supports the star against gravitational forces. A rough analogy is the motion of cars in a parking lot. When the lot is not crowded, all the cars can be in low-energy states, that is, parked. When the lot is full, some cars must be in higher-energy states, that is, moving around as their drivers search for a space.

If the star is somewhat more massive than the sun, it may undergo a supernova explosion that leaves behind a neutron star. In a neutron star, the density of particles is so large that the electrons have combined with the protons to form a star that is composed mostly of neutrons. The gravitational forces are held in check by neutron degeneracy forces. A sample of neutron-star material the size of a sugar cube would weigh one billion tons.

If the collapsing core of the star has a mass greater than about three solar masses, gravitational forces overwhelm even the neutron degeneracy forces; the star collapses in on itself to form a warp in space called a black hole. A black hole does not have a surface in the normal sense of the word. It is more like a whirlpool. It has a critical range of influence; anything that falls within that range is pulled in with no hope of escape. This critical distance from a black hole is called the gravitational horizon.

For a black hole about ten times the mass of the sun, the gravitational horizon has a radius of about 10 kilometers around the black hole. Combined radio, optical, and x-ray observations have provided strong evidence for the existence of two black holes about this size. One is called Cygnus X-1, because it is the strongest x-ray source in the constellation of Cygnus. The other is called LMC X-3, because it is the third strongest x-ray source in the Large Magellanic Cloud, a nearby galaxy.

In recent years, evidence has been accumulating for the existence of a much larger black hole in the nucleus of our own galaxy, the Milky Way. The orbital motions of clouds there imply a concentration of dark matter having the mass of 5 million suns; radio and unfrared observations suggest that this mass is confined to a very small region. In 1983 radio astronomers K. Lo and M. Claussen of Cal Tech used the VLA to show that streams of matter are converging on the center of the galaxy. The observed velocities in the gas streams indicate that matter is spiraling into a massive black hole in the center of the galaxy.

On an even larger scale, optical observations of the giant elliptical galaxy M87 in the constellation of Virgo have been interpreted to mean that a black hole having a mass a billion times that of the sun exists in the nucleus of that galaxy. The gravitational radius of such a black hole would define a sphere about the size of the solar system. And that is about the size of the powerhouses that seem to be required for quasars and violently active galaxies.

A massive black hole in isolation cannot produce the energy observed to flow out from active galaxies and quasars. It must have "food" in the form of gas. This gas could be supplied by the millions of stars that are in orbit around the black hole, or it could come from gas falling into the galaxy from outside. Wherever it comes from, the gas will probably form a disk or whirlpool, much as water swirls around a drain. As the gas gets closer to the black hole, it swirls faster; frictional processes heat it hundreds of thousands of degrees Celsius, up to millions of degrees, producing the observed optical and x-radiation.

If the pressure in this radiation exceeds a certain critical value, the inner edge of the disk will bloat up and form steep walls around the black hole. Radiation pressure and electromagnetic fields generated in the turbulent gas could lead to huge flares and the ejection of gas. The steep inner walls of the disk could funnel the gas into narrow jets, such as those observed with the VLA and other radio telescopes.

This black hole model explains why most galactic nuclei are active to some degree: they all presumably contain massive black holes. Some nuclei will have less massive ones and limited supplies of gas; these are more like our galaxy and our neighbor, the Andromeda Galaxy. Others will have supermassive black holes and large gas supplies; these are the quasars.

One of the more spectacular results obtained with the VLA is the discovery of zigzag jets. These emerge from the center of a galaxy in opposite directions, then show several sharp bends, like a corkscrew. This shape would be produced if the rotational axis of the black hole were wobbling. (Think of a water hose that is slowly rotated as the water shoots out.)

What would cause a black hole to wobble? One possibility is that another massive black hole is in orbit around it. A number of galaxies have been observed to have double nuclei, thought to be produced by a process called galactic cannibalism. In this process, a smaller "missionary" galaxy collides with a larger "cannibal" galaxy and is swallowed up. All that would survive of the missionary galaxy would be the central black hole, which would eventually go into gravitational orbit around the black hole of the cannibal galaxy. The lesser, but still considerable, mass of the missionary black hole would cause the wobbling of the cannibal black hole and hence the corkscrew jets. So far only a small fraction of radio galaxies and quasars have been mapped with the VLA, so these ideas remain largely untested. Still, the radio maps have given us important clues and stimulated new lines of thought that may ultimately lead to an understanding of these most energetic objects in the universe.

At present, black holes are enjoying immense popularity among astrophysicists. Black holes having as much mass as a billion stars and as little mass as a small mountain have been invoked to explain phenomena ranging from quasars and active galaxies to x-ray stars and the mystery of the dark matter (see Part II, below). In the next few years, data from powerful telescopes such as the VLA should tell us whether or not the popularity of black holes is justified. In the meantime, are other ideas equally as bizarre and useful as black holes being neglected, just because they are unfashionable? After all, the concept of black holes was first broached by Robert Oppenheimer and Harlan Snyder twenty-five years before it was given serious consideration by astronomers. In this connection, we asked Heeschen how much the reviewing committees at the VLA enforce their own view of reality on the observers. Do they allow time for unorthodox, off-the-wall ideas to be pursued?

"We are supposed to consciously allow for them," he said. "You should ask them what they do now at the VLA . . . I am out of the loop, so my direct involvement ended five years ago. I know the people there, and I know the general philosophy that I tried to instill for seventeen years when I was director. I am quite sure that off-the-wall experiments will get done, especially if they are by a guy that is recognized as being very competent. An off-the-wall experiment by an unknown person, unfortunately, may not fare so well. *Sometimes* unfortunately. And I think it is true that any system where you have referees tends to go down the beaten path. We may not get away from it as much as we really ought to."

Barry Clark, who now directs the observing program at the VLA, answered the same question. "That sort of thing is pretty hard to judge," he said. "We don't get very many proposals like that. But there are ones that we put on even though we really don't expect a result. A few percent or less of the available time is given to these long-shots."

We asked Heeschen which problems he would especially like to see the VLA attack. "Some of the things that I first thought about doing with it, almost twenty years ago, have not been done yet. These mostly relate to utilizing the array in its most sensitive mode. This requires 8–10 hours of observing on a given field. This isn't being done very often because of the other demands on the telescope. I think that we have yet to really understand what the radio universe looks like at these low levels of sensitivity (or equivalently in the distant past, 10 or more billion years ago, before galaxies and quasars existed). Eventually the VLA will do that and more. But it will be a while. The original concept of the array was to make such detailed maps. What has in fact happened is that the telescope is so sensitive that a large fraction of the observations are done in the so-called snapshot mode where instead of 8 hours, 5 minutes are used, so a hundred sources can be studied in one day."

Is the making of detailed maps of the radio universe at low levels of sensitivity one of his projects? "No, no. I hardly ever use the VLA. I'm getting old and there is too much involved. It is very computer-intensive and most of the programs involve a lot of people. I prefer to work alone. I do some work at the VLA, but I have other interests that are better served by other telescopes."

On the other hand, almost all astronomers feel that the interests of the astronomical community have been well served by David Heeschen in his two decades of dedication to the dream and reality of the Very Large Array. So well served that they wanted him to do it again. When it came time to find a director for the Space Telescope program, one name came up again and again. They offered David Heeschen the job. Though still an active researcher, he is happy to be free of the demands of administration, happy to have more time with Eloise, his wife of 33 years—time to hike and birdwatch in the Blue Ridge Mountains near Charlottesville and to sail up and down the eastern coastline in their 34-foot schooner. He refused the offer, politely.

_TWO

A High-Energy Astrophysicist and the Einstein X-Ray Observatory

6

When David Heeschen turned it down, the committee that was searching for a director of the Space Telescope Science Institute turned to Riccardo Giacconi. Like Heeschen, Giacconi had made his reputation in a field outside of traditional, optical astronomy; and also like Heeschen, he had spent the better part of two decades presiding over the creation of a new telescope that revolutionized the field. But apart from these similarities, the two men are a study in contrasts.

Heeschen is a soft-spoken, low-key, almost hermitic figure who never went looking for the job of director of the Very Large Array. It just came to him, because he was in the right place at the right time and the "kingmakers" were confident he could do the job. And almost at the first opportunity, which turned out to be seventeen years later, he turned the reins of power over to someone else. Giacconi, on the other hand, had been an aggressive, articulate, and outspoken advocate of an orbiting x-ray telescope since the early sixties. For over a decade he fought and won a battle to convince the scientific community and NASA that an x-ray telescope was both feasible and desirable. Along the way he and a closely knit group of hard-driving scientists and engineers organized by Giacconi designed, developed, and flew NASA's highly successful *Uhuru* x-ray satellite, the instrument that provided the first solid evidence of black holes.

With the successes of *Uhuru* and the *Einstein* x-ray telescope observatory, which was launched in 1978, astronomers have come to recognize that "x-ray vision" is an essential tool for investigating the behavior of matter under extreme conditions of density and temperature, from the crush of black holes to the vastness of intergalactic space. They have also come to recognize Giacconi, who is now the director of the Space Telescope Science Institute, as one of the dominant personalities in astronomy today. Unlike Heeschen, Giacconi has not grown uncomfortable with the rewards and responsibilities of power. If anything, the mantle fits too well. Some astronomers who find his style bruising refer to him as "Prince Machiavelli" and "General Riccardo." Others endorse it. "Sure, he's a tiger," a senior statesman among astronomers observed, "but that's what it takes to put the show on the road, so more power to him." The idea of more power to Giacconi raises the hackles of many of his colleagues, but none of them can deny that he put the x-ray astronomy show on the road.

Giacconi was born in 1931 in Genoa, Italy. His mother and father separated when Giacconi was very young, the marriage was annulled, and his mother remarried Antonio Giacconi. Riccardo remembers his stepfather fondly: "He was the one I really talked to, from the age of six on. He didn't have a great deal of culture, he was trained to be an accountant or something equivalent. He was socialist, an antifascist. He never took the card of the Fascist party, and had to live in France for a while. When he returned to Italy in 1937 or 1938, he couldn't get a job because he didn't have the card. The insurance company he had worked for would not accept anybody who was not a card-carrying Fascist." Giacconi's stepfather and mother eventually separated, but the father–son relationship did not end.

"I kept talking to him," Giacconi recalled, "and I think he had a very profound influence." They would take long walks together and have discussions about everything from the way energy was stored in batteries, to human nature, to the future of Italy. "I remember when the war started. I was seven or eight years old, and there was great rejoicing in the streets. I came home all excited and I said, 'Daddy, daddy, war's started, isn't it great?' and so forth. He slapped me across the face."

Giacconi picked up his pipe and lit it. Apart from a fountain pen, the type that fits into a stand, an appointment calendar, a can containing several sharp pencils, and two neatly stacked pads, his pipe and lighter were the only items on his desk—which was not a desk at all but a long white table. The carpeted corner office was tastefully decorated in grays, whites, and browns, with large potted plants and expansive windows overlooking a small park with a tree-lined, bubbling brook.

Giacconi's taste for neat, attractive workspaces created a minor flap when he moved his x-ray astronomy group from American Science and Engineering, a small Cambridge company, to the Harvard–Smithsonian Observatory. The group took over the fourth floor of the observatory complex, but not until, at Giacconi's insistence, the khaki-brown walls were painted white. This act, which Giacconi maintains was taken only for the purpose of brightening up a drab environment, was widely interpreted by the old guard at the observatory as a symbol that a new guard had arrived. Underground memoranda were circulated, warning, in satirical terms, of the impending takeover of the observatory by Giacconi's new High Energy Astrophysics Division. Giacconi tried to allay these fears by pointing out that he had made the move to the Harvard–Smithsonian Observatory out of respect for its diverse group of astronomers, and had no intention of engineering a "palace coup." Nevertheless, because of the large contracts they were able to bring in, the new division did quickly establish its dominance at the observatory. Though many scientists in other divisions now routinely join in collaborations with members of the High Energy Astrophysics Division in writing scientific papers, a lingering distrust and fear still remains in the minds of more than one of the scientists in the other groups, years after Giacconi's departure.

Giacconi continued talking about his stepfather. "He was always the odd man out. He had an incredible perception for 'the king is naked' type of thing. He had an understanding, a vision of reality which was deeper than I have ever been able to develop myself. He was much brighter than I was, I think, in understanding the world. In difficult times I go back to him, to his memory, and ask myself, what would my father do?"

Giacconi's mother also strongly influenced both his character and his career. She was a teacher of mathematics and physics at the high school level, one of the first women in Italy to get a college degree, and a successful author of mathematics textbooks. Though she is over eighty years old, she continues to revise her books and write new ones. "I never learned any mathematics from her because I detested working with her," Giacconi said. "I learned it by myself, without telling anybody I knew about it. But she knew. She had this trick. She would ask me to help her. For instance she would bring home tests, written tests that she had given in school, and she would ask me to please help her correct them. And so I would correct the test." He laughed as he remembered this little game he and his mother used to play with each other.

"She was totally indifferent to politics. She only cared about the family. I mean, me. She only cared about me. Period." He laughed again. "And the only other thing, which she denies, but I believe was important, was that I had to succeed. There was no question that I had to succeed. For example, I would come home from an exam with a grade of 27 out of 30, and she said, 'How come you did so poorly?' After that I knew that I had to get 30 out of 30 or forget it. I think she transferred a lot of the frustration she felt, being a bright woman in Italy, to a determination that I should succeed."

We asked Giacconi how he came to be a scientist. "In my last year in high school I was interested in many things," he said. "In particular, I was interested in art, architecture, philosophy, theology. The things that really attracted me was to go around museums and to read books on architecture. I became really interested in architecture. What stopped me there was that I thought about my ability to create—I had big dreams about architecture. If I was going to be an architect, I was going to create shapes never before seen by man, I mean a new style or something. And I was very worried about my ability to create, that I couldn't be creative enough to make it. So I decided no, I couldn't do that. So I considered philosophy."

"I would stay up nights to read philosophy. At that time I hadn't completely shaken off the Catholic period of my life, which ended when I was 17. So I was reading theology, and metaphysics. My favorite author was Kierkegaard. Then I started thinking about career choices. I could see my mother's life as a teacher and I didn't want that life. My mother was pushing for a more practical career. She decided that I was managerially oriented, and not a serious student, so I should take engineering. In Italy, if you get very high marks in the fourth year, you are allowed to jump a grade and take exams which allow you to go to the university. That's what I decided to do, because I was sick and tired of high school and wanted to go to a university."

"I passed the exams with flying colors, every one except physics, in which I made a minimum passing grade. I did great in history of art, architecture, and so forth. So, I decided that, well, nuclear power was around, so there might be some opportunity there. I decided to take physics, because it wasn't as bad and dull as engineering. If one was creative, one could do something. And if one was mediocre, one could have a job. So it was just at that level that I chose physics. Because it wasn't architecture, with which I was passionately in love, and it wasn't philosophy, which I really thought was important, but it was okay."

What about astronomy? Was that a childhood interest? "No. I had a friend, who still is an amateur astronomer, who used to build telescopes; he would invite me to his house and we would peer through lenses and so on. I thought he was a bit crazy. In fact, I thought astronomy was a bit like zoology. It wasn't a serious science, whereas physics was. It was good stuff."

At the University of Milan, Giacconi began working in the physics laboratory, building power supplies, doing literature searches, and eventually conducting experiments on his own. By 1954 he had received a Ph.D. in physics, with a specialty in cosmic ray physics. Cosmic rays are the energetic nuclei of atoms—for the most part hydrogen atoms—that constantly bombard the top of earth's atmosphere. They arrive from all directions; this indicates that they do not come from the sun, but from the stars or from interstellar or intergalactic space. In the years before high-energy particle accelerators were developed, the only way that physicists could study the interactions of nuclear particles at high energies was to study the interaction of cosmic rays with the atoms of the atmosphere. The competing theories predicted different types of interactions that produced different types of short-lived elementary particles. It is the difficult, often frustrating task of cosmic ray physicists to design experiments to detect these evanescent particles. For example, Giacconi had spent two years collecting evidence for only 80 cosmic ray interactions for his thesis work.

After completing his thesis and receiving his degree, Giacconi remained at the University of Milan for two years as an assistant professor. In 1956 he came to the United States, to Indiana University, on a Fulbright scholarship. It was there that he married Mirella Manaira, a bright, attractive Milanese woman whom Giacconi had known since his student days in Italy, and who would be a source of strength and a valued confidant in what she called the whirlwind years of the next two decades. On the professional side, though, the two-year stay at Indiana University was for the most part fruitless. He moved next to Princeton University, where he took a position as a research associate at the Cosmic Ray Laboratory.

With Herbert Gursky and Fred Handel, Giacconi put together experiments to search for hypothetical particles produced by the collisions of high-energy cosmic rays with atoms in the atmosphere. "It was wonderful training," he recalled, "I learned a lot of technology."

But from a purely scientific point of view, it was unproductive. They did not find the hypothetical particle, or anything else of particular interest. Giacconi felt he was wasting his time. The usefulness of cosmic ray research as a tool to study high-energy nuclear collisions was rapidly being surpassed by the development of high-energy particle accelerators. But research with accelerators was done by large groups, and, as Giacconi put it, "I wasn't really that eager to go into large groups. I wasn't doing very well. Nothing I had done in science was really of great significance. I was on this fellowship, you know, going from one fellowship to another. I didn't have a permanent position in Italy. There wasn't very much for me to do . . . I was on a visitor visa and I had to have it changed. Princeton didn't feel like they could change it for me."

The solution to Giacconi's dilemma came through his network of colleagues. In the course of joint experiments with MIT, he met Herbert Bridge, an MIT professor. Bridge introduced Giacconi to Martin Annis, a recent graduate who was the president of American Science and Engineering (AS&E), a company he had helped form less than a year before. The National Aeronautics and Space Administration (NASA) had just begun operations, and Annis was looking for somebody to start up a program in space science at AS&E. They offered Giacconi the job, for $13,000 a year. "I couldn't wait to accept it," Giacconi recalled with laughter. "I was afraid they might change their minds. I couldn't believe that somebody actually wanted me, or would pay for my services."

7

When Giacconi arrived at AS&E, he knew that he wanted a change from cosmic ray research, and he knew that Annis wanted him to start a program of some sort in space science; beyond that he had no plan in mind. For one thing, he knew little about what was going on in space research. "I first knew about Van Allen in '56 [James Van Allen, the discoverer of the Van Allen radiation belts in the upper atmosphere]. And frankly, it seemed to me that serious work was in elementary particles, not this fooling around with rockets to look at auroras. But I had the opportunity to do anything I wanted. They [AS&E] didn't have a program, and the question was to figure out a program and get it started."

For the first few months, Giacconi explored the same sort of problems that other fledgling space-science groups were attacking at the time. He designed experiments to measure the properties of charged particles in the Van Allen radiation belts and he began studying ways to build a directional detector to measure gamma rays emitted in a nuclear bomb explosion in space. Then, in September of 1959, he was invited to a party at the home of Nora and Bruno Rossi.

Bruno Rossi, chairman of the board of AS&E and professor of physics at MIT, was one of the most distinguished experimental physicists of the time. He had written classic books on cosmic rays and optics. Like Giacconi, he had emigrated from his native Italy to the United States, though, unlike Giacconi, he came not so much in search of new opportunities as in search of a haven from Fascism. He had worked on the Manhattan Project, which produced the first atomic bomb, and had done pioneering research in cosmic ray physics. When the National Academy of Sciences established a Space Science Board in 1958, Bruno Rossi was asked to serve. The Space Science Board was a group of eminent scientists who were to help NASA formulate a broad policy on space research.

To assist them in their work, the Space Science Board drew upon the advice of a number of subcommittees. Three committee members—John Simpson of the University of Chicago, Leo Goldberg of Harvard University, and Laurence Aller of the University of California—suggested that a survey of the sky with x-ray detectors might prove interesting. Because the earth's atmosphere absorbs x-rays, x-ray astronomy can only be done from space. The detectors have to be launched about a hundred miles or more above the surface of the earth. In the late 1940s and early 1950s, Herbert Friedman and his colleagues at the Naval Research Laboratory had used x-ray detectors aboard V-2 and Aerobee rockets to show that the sun was a source of x-radiation and that x-ray observations of the sun were essential for understanding energetic events in the solar atmosphere, such as solar flares. But Friedman's successful x-ray studies of the sun had not stimulated other scientists to enter the field and to attempt to detect x-rays from other stars. For one thing, the field was too risky. The rocket could blow up on the launch pad, or the experiment could fail for any one of a hundred reasons once it was launched, in which case there was no opportunity to fix it. And even if all went well, the experiment could fail simply because there were no detectable x-ray sources. The x-ray power of the sun is a million times weaker than its power at optical wavelengths. If the sun was typical, the astronomers would need detectors that were a thousand times more sensitive than the ones already in existence to detect x-rays from other stars. So, for a variety of reasons, astronomers in the 1950s shied away from x-ray astronomy and left Friedman and his group to pioneer the field. This attitude still prevailed in the late 50's, and most astronomers ignored the suggestions of Simpson, Aller, and Goldberg. But a few scientists, mostly nuclear or cosmic ray physicists looking for new fields of research, listened. Among them was Bruno Rossi. He was not discouraged by the inherent difficulties of doing space astronomy and the pessimistic predictions that cosmic x-ray sources would be undetectably weak. Rather, he felt that since virtually nothing was known, there was a possibility for major new discoveries.

At the party, Giacconi was introduced to Rossi. After exchanging pleasantries about the relative merits of Italy and New England, Rossi suggested to Giacconi that he should use his position as director of the Space Research Division at AS&E to develop a program in x-ray astronomy. It was, Giacconi recalled, "a seminal suggestion, which all of a sudden gave me a way to go. It was terribly important."

A suggestion was one thing; making it bear fruit was quite another. For the next ten years, Giacconi worked, by his own account, "as a man possessed." At the start he had a small group of half a dozen or so engineers, technicians, and scientists working in an old milk-truck garage that had been converted into a laboratory. A decade later, the Space Research Division had eight converted milk-truck garages for laboratories and over a hundred employees. "It was a hell of a lot of work, with lots of failures along the way."

Giacconi's experience in cosmic ray physics had convinced him of the frustrations of doing research in a field where the "events" or particle detections were few and far between. The early estimates of the numbers of x-rays from cosmic sources indicated that they too might be scarce. There are two solutions to this problem. One is to build huge detectors, which are basically gas-filled Geiger-type counters, so as to collect more x-rays. They work like a window: the larger you make it, the more light you let in. This approach was pursued by Friedman's group at NRL, and led eventually to the large-area x-ray detector which flew as part of NASA's first High Energy Astronomy Observatory. Such detectors are severely restricted in their efficiency by "background noise." As the detector area is increased, the amount of background noise from cosmic rays, charged particles in the Van Allen belts, or x-rays from distant background sources also increases. The net effect is that the sensitivity of large detectors of this type increases only slowly, as the square root of the size of the detector. An x-ray telescope, by contrast, has a large collecting mirror which concentrates x-rays from a small region of the sky onto a small detector. If the collecting area is large enough and the focusing capability of the mirror is good enough, then the background noise can be reduced to almost negligible values. The sensitivity of an x-ray telescope can easily be a million times greater than a large window-type detector.

But x-rays do not reflect from mirrors the same way that light waves do. Because of their short wavelength and high energy, x-rays penetrate into the mirror and are absorbed by it, in much the same way that a stream of bullets would hit a wooden wall. However, just as bullets ricochet when they hit the wall at a grazing angle, so too will x-rays ricochet off mirrors. Giacconi realized that this effect could be used to make a mirror that could focus x-rays. Hans Wolter, a German physicist who had attempted to make an x-ray microscope, had shown that x-ray images could be formed if the incoming x-rays were made to reflect twice from properly shaped mirrors. But it was too difficult to construct mirrors of the required precision on the small scale required for microscopy, so Wolter dropped the project.

Giacconi recognized that Wolter's ideas might work on the larger scale of telescopes. The shapes of these bigger x-ray mirrors should not be much more difficult to grind than those of optical telescopes. He discussed the idea of an x-ray telescope with Martin Annis, who in turn told Bruno Rossi. Rossi responded eagerly. He pointed out that, since the x-rays would be reflecting off the mirrors at very shallow angles—that is, they would be coming in almost sideways rather than straight on—one or two mirrors could be nested one inside the other in concentric cylinders. In this way, the effective collecting area could be doubled or tripled. Giacconi and Rossi published a description of an x-ray telescope, and Giacconi immediately set to work to build one. It was 1960.

Giacconi realized that the construction and flight of an x-ray telescope was years away, so he began a parallel effort to build better conventional detectors in order to detect x-radiation from stars. The proposal he submitted to NASA for this purpose was rejected. Undeterred, Giacconi surveyed his other options for funding. AS&E was at that time doing classified defense work for the Air Force Cambridge Research Laboratories (AFCRL), a research facility located 20 miles west of Boston. Since AFCRL was known to be interested in studies of the sun, Giacconi approached them with a modified x-ray astronomy proposal: AS&E would study solar x-ray emission, as well as x-ray emission from the moon. He argued that high-energy particles streaming away from the sun might produce x-rays when they struck the moon. Early in 1960 AFCRL agreed to fund the research.

At the time, Giacconi's team consisted of an electronic technician, a mechanical engineer, and two physicists, Frank Paolini and Norman Harmon. By the following June this group had put together an experiment to be flown on a small rocket. Unfortunately, the Nike-Asp rocket engine failed. A second proposal to AFCRL was accepted and another flight was scheduled for October 1961. In the meantime, Herbert Gursky, with whom Giacconi had worked at Princeton, joined the group. By October they were ready, with a detector a hundred times more sensitive than any that Friedman's group at the Naval Research Laboratory had flown. Full of confidence, they traveled to White Sands Missile Range in New Mexico, where an Aerobee rocket was waiting to carry their experiment 200 kilometers above the surface of the earth.

Once again, they met disappointment. The rocket launched successfully, but its door stuck. Their detectors saw nothing but the inside of the rocket. A determined Giacconi scheduled another launch for eight months later, in June. One minute before midnight on June 18, 1962, at White Sands, they launched their third try at detecting x-rays from beyond the solar system. The rocket was above 80 kilometers for a total of 5 minutes and 50 seconds, and reached a maximum altitude of 225 kilometers. As the rapidly spinning rocket climbed above 80 kilometers, the rocket doors opened, giving their detectors their first view from space. Each time the counters spun past a certain point in the southern sky, they registered a significant increase in the rate of detected x-rays. Giacconi's group had discovered a cosmic x-ray source that was hundreds of times brighter than anything anyone had suspected would exist.

For the next two months they analyzed the data from the rocket flight to make sure that they had not been fooled by ultraviolet radiation leaking into the detectors, or by charged particles from the Van Allen radiation belts, or by some mechanical malfunction. In late August, at the Third International Symposium on X-ray Analysis at Stanford University, Giacconi, Gursky, Paolini, and Rossi were sufficiently confident of their results to announce the discovery of the first cosmic x-ray source, Scorpius X-1 (so called because it was the strongest x-ray source in the constellation of Scorpius). In the audience sat Herb Friedman. When he heard Giacconi describe the strength of the source, he realized that, with a little more effort, his group could have detected the same source several years earlier. Instead, they had chosen to shift their attention to ultraviolet astronomy, which many astronomers had suggested would be more promising. Friedman introduced himself to Giacconi and congratulated him on a major new discovery. He then returned to his laboratory to galvanize his group into action. By April of 1963, Friedman's group had provided independent verification of the AS&E discovery and had detected another strong source associated with the Crab Nebula, the remnant of a supernova explosion. Even the skeptics had to admit that a new field of astronomy had been born.

For the astronomical community, it was as though gold had been discovered in the Klondike. Most members expressed some interest, and then continued to go about their business as usual. Some ignored the discovery altogether. But a few opportunists rushed to get into the field, to stake their claims, and, they hoped, to open up rich new veins of research. Giacconi, with the help of Herb Gursky and others, immediately began to develop a long-range plan. "At that time only the existence of Sco X-1 and a source in the Crab Nebula were known," Giacconi recalled. "In spite of this, I was young and enthusiastic enough to present to Nancy Roman [chief of the Astronomy Branch of NASA] a long-range program for x-ray astronomy observations for the next decade."

The first phase of his audacious program was to continue rocket surveys of the sky for new sources, to study selected sources in detail, and to test new instrumentation. The second phase would be to place an x-ray experiment on the fourth Orbiting Solar Observatory. These observatories were large, spinning, wheel-like spacecraft whose axis pointed toward the sun. The major experiments were positioned along the axis and secondary experiments were put in the rim of the wheel; instruments in the rim could be used to scan the sky as the wheel rotated. The third phase of the program would be a satellite designed exclusively for x-ray astronomy observations. The fourth phase would be a modest-sized spacecraft carrying a focusing x-ray telescope. The fifth phase would be a full-fledged large x-ray telescope. It was projected to fly in the late sixties.

"It is important once in a while to step back and say what would I really like to do?" Giacconi said, "and I think we did that pretty well. Maybe it was the atmosphere of a small private company like AS&E, maybe it was the group . . . we had this feeling of machismo that we could do whatever was needed."

8

Realistically, Giacconi expected to get NASA approval for the rocket survey program, and possibly a space on the Orbiting Solar Observatory. So when AS&E learned that NASA was also definitely interested in pursuing the idea of a satellite devoted exclusively to x-ray astronomy, the group was elated. Within months, AS&E submitted to NASA a detailed proposal for an "x-ray explorer satellite" to be launched sometime in December of 1965.

A NASA review committee of astronomers, chaired by Nancy Roman, endorsed the mission. But not for a flight in 1965 or 1966 or 1967. As it turned out, the X-ray Explorer was not launched until December of 1970, some five years after the proposed launch date. The delay was due in part to the glacial slowness with which the NASA bureaucracy moves, and in part, paradoxically, to NASA's enthusiasm for the idea of small satellites devoted exclusively to one instrument. It was such a good idea, NASA decided, that it should be generalized from one project into an entire program of "Small Astronomy Satellites," of which the X-ray Explorer would be the first. This new plan raised the ante. A second round of approvals was needed and a whole new level of management had to be put in place.

In the meantime, Giacconi continued to push forward on other fronts. AS&E's active rocket program, this time supported fully by NASA, kept them at the rapidly expanding frontier of x-ray astronomy and allowed them to test new instruments and techniques which would be incorporated into the X-ray Explorer Satellite. Not content to be ready for the satellite, which would represent a quantum leap in the state of the field, Giacconi labored, both in the laboratory and in the halls of NASA, to keep two moves ahead and to ensure that an x-ray telescope observatory would be flown in the foreseeable future—and that it would be built and directed by his group.

Using funds from NASA, they set up a laboratory dedicated to the development of an x-ray telescope. They machined and polished the interior surfaces of aluminum tubes and coated these surfaces with gold to make them highly reflective. By 1963 they had constructed an x-ray telescope with a collecting area that was roughly equal to that of a dime. "Even at this early stage of development," Giacconi said, "it was clear—to me at least, if not to NASA, that grazing-incidence telescopes would be quite useful in the x-ray study of celestial objects, particularly the sun." More funding was needed to build a telescope with precision polished surfaces, and to develop new techniques for preparing the peculiar-shaped surfaces needed for x-ray telescopes. But funds of the magnitude needed were not likely to come out of NASA's relatively small laboratory research budget. What was needed was NASA approval of an x-ray telescope for an actual flight mission; to ensure success for the mission, ample funds would have to be budgeted for research and development. But NASA still felt that AS&E's x-ray telescopes were too crude, and that there was no guarantee they would work under flight conditions. So it appeared that the x-ray telescope project was stuck at square one. NASA refused to provide sufficient funding until the telescope technology was improved, and the technology could not improve until NASA provided sufficient funding.

What was needed was someone with faith in the project and the willingness to take a risk. Although the AS&E management did take risks and provide corporate funds for the development of promising projects, these risks involved hundreds of thousands of dollars at most, not millions. They simply did not have the money. Someone at NASA would have to stick his neck out. Fortunately, someone did. John Lindsay had been a member of Friedman's solar x-ray astronomy group at the Naval Research Laboratory. Shortly after the establishment of NASA, he had moved to NASA's Goddard Space Flight Center in Greenbelt, Maryland, where he had provided the initial stimulus for the Orbiting Solar Observatory satellites. He had also been the technical monitor for the NASA contract that had allowed AS&E to start the development of x-ray telescopes. Because he was personally interested in solar x-ray astronomy and because he was impressed by the work at AS&E, he used his influence to secure funding from NASA's Solar Physics Branch for the development of x-ray telescopes.

Lindsay and his colleagues at Goddard joined Giacconi and his colleagues at AS&E in a joint venture. The result was an 80-centimeter-long x-ray telescope that looked something like a mailing tube. With this telescope the first telescopic x-ray images of the sun were obtained in October 1963, and others in March 1965. A much bigger telescope was scheduled to be part of a manned spacecraft. Then, in September of 1965, John Lindsay, an apparently healthy man in his forties, suffered a fatal heart attack while mowing his lawn.

"The AS&E–Goddard collaboration was dissolved after Lindsay's death," Giacconi said, "and the groups went their separate ways." But Lindsay's efforts were not in vain. Eight years later, two x-ray telescopes—one built by AS&E, the other by Goddard—would send spectacular x-ray pictures of the sun back from *Skylab*. According to Giacconi, "The debt that these achievements owe to John Lindsay's advocacy cannot be overestimated."

Despite the successful solar x-ray telescope flights in 1963 and 1965, and the strong support of Harold Glaser of NASA's Solar Physics Branch, the idea of an x-ray telescope devoted to the study of objects outside the solar system still met strong resistance. For one thing, NASA officials saw no urgent need for x-ray telescopes. X-ray stars were so bright that a lot of good science could be done with conventional x-ray detectors. Also, there was still some skepticism within NASA as to whether x-ray telescopes would work on the extrasolar sources (sources outside the solar system) which, though bright, were nevertheless millions of times less intense than x-rays from the sun because of their enormous distances. The mirrors would have to be polished to a precision comparable to that of the finest optical telescopes. Furthermore, the solar experiments had used film to record the incoming x-rays, which was retrieved at the end of the rocket flight or, in the case of *Skylab*, by the astronaut. For an x-ray telescope aboard an unmanned satellite, an electronic x-ray imaging camera that could detect individual x-rays and record them electronically would be needed. Such a camera would have to be developed. In view of these considerations, NASA felt in the mid-1960s that an expensive program to develop a sophisticated x-ray telescope was premature. The scientific community agreed, for the most part. There was a strong feeling that they should proceed with caution, a feeling that more all-sky surveys with conventional techniques were needed to see if x-ray telescopes were really necessary. After all, what if x-ray astronomy were to turn out to be the study of a few peculiar stars that are of no general interest to anyone except specialists? "What's the rush?" seemed to be the predominant mood among the scientists as well as NASA.

Giacconi was undaunted. Even though NASA would not support a laboratory program for the development of an extrasolar x-ray telescope, they did have a contract for the construction of a solar x-ray telescope. "We bootlegged the developments necessary for an extrasolar telescope as part of the solar program," Giacconi said. "A two-nested mirror configuration was used, although it was not strictly necessary, and first generation x-ray television cameras were developed although film was used." In this way, the AS&E group kept the technology necessary for an extrasolar telescope moving forward.

"The lack of support for use of focusing optics by the scientific community was a more serious problem," Giacconi recalled. To overcome it, he assiduously talked up the potential scientific value of an x-ray telescope wherever and whenever he could. He understood, as do all successful practitioners of the politics of science, that if he wished to see his dream become a reality, he had to have support for his idea among scientists at the grassroots level. He had to persuade them that an x-ray telescope would be a great thing for astronomy, and that he could build one. Then, maybe, they would lend their support to the concept on the influential advisory committees. Giacconi argued his case through personal contacts, through lectures at universities, laboratories, and scientific meetings, and through appearances whenever possible before scientific advisory committees such as the Astronomy Missions Board of NASA and the Space Science Board of the National Academy of Sciences.

Thus it was that Giacconi went to Woods Hole, Massachusetts, in the summer of 1965, to participate in a meeting of the X-ray and Gamma Ray Astronomy Panel of the Space Science Board. The panel had convened to study plans for future x-ray and gamma ray experiments. Herb Friedman was chairman; Bruno Rossi and George Clark from MIT were there, as were many other influential scientists, including Frank McDonald, a cosmic ray physicist from Goddard Space Flight Center. McDonald and others were especially interested in putting together a program that would use a series of identical and relatively inexpensive spacecraft built from left-over *Apollo* program hardware.

"To me it was a tremendously important meeting," Giacconi recalled, "because I was given an opportunity, for the first time, to really try to debate publicly why we should have x-ray telescopes.

I went there grim and determined that I would do it, I would explain what really had to be done." He succeeded, at least partially. The panel agreed that x-ray telescopes would be necessary, once the field progressed beyond the exploratory stage, and recommended that a program to develop an x-ray telescope be started as soon as possible. They also recommended that the extremely large, conventional x-ray detectors promoted by Herb Friedman be flown prior to the x-ray telescope, as well as a large cosmic ray experiment proposed by Frank McDonald.

This recommendation resulted in an alliance between the two dominant figures in x-ray astronomy, Giacconi and Friedman, and a highly respected and influential NASA scientist, McDonald. Together and individually they continued to campaign for the general idea of large spacecrafts for astronomical observatories and for their own particular ideas as to what form these observatories should take. McDonald was particularly effective within NASA. In 1967 NASA established the Astronomy Missions Board to serve in an advisory capacity in the planning of astronomical experiments in space. The board was composed of 18 astronomers, including Friedman and three other scientists who were actively involved in high-energy astronomy: William Kraushaar of the University of Wisconsin, Laurence Peterson of the University of California, San Diego, and John Simpson of the University of Chicago. Over several months, the board heard presentations from supporters of a wide variety of competing missions, as well as reports from the 31 other scientists who had been recruited to advise the board members. In the end, they recommended a greatly expanded program in high-energy astronomy. Top priority was awarded to a series of missions aboard large spacecraft, called High Energy Astronomy Observatories. One of these spacecraft, they recommended, should be an x-ray telescope observatory.

The recommendations of the Astronomy Missions Board, though of no official weight, were crucial. They showed that a broad spectrum of scientists approved the concept of large spacecraft dedicated to x-ray, gamma ray, and cosmic ray experiments in general, and that they approved the x-ray telescope in particular. This support from the scientific community made it possible for Jesse Mitchell and Richard Halpern, the advocates of the High Energy Astronomy Observatory program within NASA, to plead their case effectively.

Funds were appropriated for a feasibility study of the High Energy Astronomy Observatory concept within NASA, and Halpern became program manager. With the help of McDonald and Alois Schardt, another NASA official, he built a strong case for the program in government circles. They found powerful allies at NASA's Marshall Space Flight Center in Huntsville, Alabama. Huntsville gained fame as the home of Werner von Braun and his team of German rocket scientists and engineers. It was there that the first rockets to launch Americans into space and toward the moon were built. By the late 1960s the peak rocket-building effort of the *Apollo* program was over, and von Braun and his associates, including Ernst Stuhlinger, were looking for another large NASA program. So when McDonald and Schardt approached them with the idea of managing the High Energy Astronomy Observatory program, they responded with enthusiasm. They assigned engineers to help in the feasibility study, and came up with a basic spacecraft design, which was essentially a large boxcar weighing about 12 tons.

More than a year passed, filled with intense lobbying within the scientific community, NASA, the Office of Management and Budget, and Congress. Finally, in early 1970, NASA issued an Announcement of Opportunity to submit proposals for experiments to be flown on one of a series of High Energy Astronomy Observatory missions. They received proposals from virtually every x-ray, gamma ray, and cosmic ray astronomy group in the United States. Giacconi and Gursky's group at AS&E joined a consortium with an MIT team led by George Clark, a Goddard Space Flight Center team led by Elihu Boldt, and a Columbia University team led by Robert Novick. In May 1970 they submitted a single proposal, with Giacconi designated as principal investigator.

In keeping with his philosophy of leaving as little as possible to chance, Giacconi had one of the AS&E vice presidents drive from Cambridge to Washington with the required 50 copies of the proposal in his station wagon and deliver them in person to ensure that they arrived before the deadline. In October of 1970 the consortium made an all-day presentation before NASA management and a reviewing committee of scientists from universities and government laboratories. Also making a presentation that day to the same reviewers was another consortium that was proposing an x-ray telescope. Both groups were allowed to sit in on the presentations of the other group and ask questions. Here again Giacconi had left nothing to chance. The consortium had practiced their presentations several times before the most critical of audiences—one another. They had made sure that all their demonstrations of detectors worked, and they had brought backup personnel to deal with esoteric points that arose in the questioning.

The well-prepared AS&E-Columbia-Goddard-MIT presentation overwhelmed their competitors. Their victory was so thorough that it led one of the reviewers to express the reservation that Giacconi and his colleagues might dominate the field and eventually exclude all competition (a charge that has surfaced time and again). At this point in the meeting Giacconi asked for the floor and delivered an eloquent statement of his goals and philosophy.

"I want to reassure everyone," he said, "that we are not barbarians from the north, coming to sack Rome. It is our intent to operate the observatory as a national facility available to all astronomers, and we will make every effort to involve other astronomers through coordinated activities using other telescopes. We also intend to turn over the full capability of the x-ray observatory to interested observers for their exclusive use for some reasonable fraction of the available time in orbit. Our purpose is not to restrict the access to x-ray data to a small coterie, but to create a facility, a national x-ray observatory, that will serve the entire community."

Giacconi's speech smoothed over some of the feathers that had been ruffled in the rough and tumble arena of the NASA reviewing room, and he lived up to his words. Over 600 guest observers from more than 50 different institutions have had a chance to accumulate and use the data from the *Einstein* x-ray observatory.

NASA officials retained some reservations, though. Despite the impressive presentation by the consortium, they were still not convinced that such an exceedingly sophisticated x-ray telescope system could be made to work in space. Out of all the proposals received for the High Energy Astronomy Observatory missions, four missions were formulated. The first would carry four x-ray and two cosmic ray experiments. The second mission would carry two x-ray experiments. The x-ray telescope observatory was planned for the third mission, and the fourth would carry an array of cosmic ray experiments. The good news to the consortium was that they had been selected for one of the four missions. The bad news was that they were third in line, and only the first two projects had final approval and funding. They were still stuck in the study phase. But Giacconi was not discouraged. "I knew that ultimately we would have an x-ray telescope. I was determined that we would have it. I knew it would work."

9

Another critical point in Giacconi's career came soon after the x-ray telescope review. In November of 1970, the X-ray Explorer experiment payload was shipped to Kenya in preparation for launch. Kenya was chosen because Giacconi and his group wanted to launch the satellite into an equatorial orbit. A rocket launched from the equator gets an extra boost of about 13 percent in speed from the higher speed of rotation at the equator compared with Cape Canaveral in Florida or Vandenberg Air Force Base in California. An equatorial orbit also avoids the troublesome Van Allen radiation belts, regions of the earth's magnetic field where the concentration of charged particles is high. The launch site was the San Marco Platform, a modified oil-drilling rig 3 miles offshore from a small Kenyan village where the twentieth century had yet to arrive. The villagers had no electricity or indoor plumbing, and still cooked over open fires. The San Marco Platform had been prepared for rocket launches by the Italian Space Agency.

An advance team from AS&E, led by Harvey Tananbaum, along with teams from the Applied Physics Laboratory (APL) of Johns Hopkins University and from NASA, had gone to Kenya with the instrument to perform prelaunch checkouts. Giacconi and Marjory Townsend, the NASA program manager for the X-ray Explorer, arrived in late November. As the launch date approached, tensions rose. Two days before launch, one of the batteries that was to supply power to the satellite began to give anomalous readings.

"The battery is failing."

"No, it's all right. The readings are spurious."

"How do you know?"

"I don't know for sure."

"Do you want to risk your career on it?"

"Do you want to pull the whole package out and put in a new battery that hasn't gone through checkout? Do you want to risk your career on a new battery working, and on someone not screwing up when they take everything out and put it back in again? We don't have time to check it all out again."

When the going got tough, the APL group, the ones responsible for the batteries, decided to get going—they took a boat back to shore and left the decision to Giacconi and Townsend. The battery, it was decided, would be replaced. Launch was rescheduled for dawn on December 12, 1970. Giacconi, unlike the APL team, did not want a single decision to be made without his approval, so he spent the night before the launch on the platform. "It was wet and cold out there," Giacconi remembered, "and I began to get chilled. I couldn't sleep. One of the Italian crew members literally gave me the shirt off his back, so I could get a few minutes' sleep and be ready for the launch. It was a beautiful gesture."

As the sun rose on December 12, Kenya's independence day, one minor delay after another stalled the countdown. As the morning wore on, the temperature on the platform began to rise. And so did the temperature inside the spacecraft, and the tempers on the platform. Giacconi was worried that the high humidity would cause oxidation of the super-thin beryllium windows on the x-ray detectors. If this happened, the windows would become brittle and might break. He wanted to launch immediately, but the APL group kept equivocating, while the Italian crew was spraying the satellite with dry nitrogen, trying to prevent oxidation.

"This heat may be too much for the electronics. Maybe we should abort and reschedule."

"No. We have to launch now. Immediately."

"What if it doesn't work?"

"It will never work if you don't get it up in the air!"

Townsend agreed with Giacconi. Shortly after noon, a Scout rocket carried the satellite into a 500-mile-high orbit over the equator. But was it working?

"I had to know," Giacconi said. "I couldn't wait until Goddard decided to turn on the instruments. I convinced Marjory Townsend to break with protocol and take a peek when the satellite made its first passage over Kenya, about an hour and a half later. We jumped in one of those rubber boats and rushed back to camp, some 3 miles away. We reached the control van just in time to turn on the high voltage to the detectors, just to see if they were working. They were, perfectly, so we shut them off again. We were like kids, we were so excited."

The excitement only increased when Giacconi returned to Cambridge to begin analyzing the data from *Uhuru* (the Swahili word for "freedom"), as they nicknamed the satellite, since it was launched on the Kenyan day of independence. "It was beautiful, almost mystical, to see the data coming in. Suddenly we had our own channel of communication with nature."

Prior to *Uhuru*, most of the information about cosmic sources of x-rays had been glimpses obtained from rocket flights of a few minutes' duration. There was no understanding as to how many sources there were, or what types of objects were producing them. It was only known that many of the sources were some peculiar type of star that gave off most of its energy in x-radiation. With *Uhuru* it became possible to monitor x-ray stars for prolonged periods of time. The result was a breakthrough of the kind that most scientists dream of but few ever experience.

It started with the first observations of Cygnus X-1, a bright x-ray star that was one of the most perplexing of a class of perplexing objects. Several different groups had measured the intensity of x-rays coming from Cygnus X-1 on several different occasions. Sometimes the results of the different groups were consistent, sometimes they were not. Had some of the observations been subject to larger errors than others, or was the intensity of x-radiation from Cygnus X-1 changing erratically with time? The discussion of these issues in public as well as private meetings was spirited, occasionally acrimonious, as insinuations of sloppy research techniques were exchanged.

In the meantime, in 1968, the Cambridge University radio astronomers Jocelyn Bell-Burnell and Anthony Hewish had discovered pulsars. Pulsars are radio sources that pulse on and off with great rapidity and astonishing regularity—some of them blink on and off with the precision of a clock that loses less than a second in a million years. The time between pulses varies from one object to the next, but is typically one second or less. The most energetic pulsar is associated with the Crab Nebula, a remnant of a supernova explosion. The regularity of the pulses can be understood only in terms of the rotation of a very large, coherent body, that is, a star. The *rapidity* of the pulsations, especially for the Crab Nebula pulsar, which pulses 30 times a second, can be explained by only one particular type of star, a neutron star. Any other star would fly apart if it were to rotate as fast as pulsars were rotating.

Since the Crab Nebula pulsar was a strong x-ray source, theorists immediately proposed that x-ray stars were also pulsars. In these models, the x-ray stars, like the Crab Nebula pulsar, were energized by a highly magnetized, rotating neutron star. A major problem for these models was that the Crab Nebula pulsar had a large, glowing nebula around it, whereas the x-ray stars did not. The defenders of the pulsar models argued that perhaps the x-ray stars were much older, so that the nebula had disappeared, or perhaps the nature of the environment or the explosion or the residual neutron star was different, so that no nebula was produced. Another problem was that none of the x-ray stars had shown evidence for periodic pulsations on a scale of seconds. But such measurements were difficult and in most cases impossible in a short rocket flight.

"With *Uhuru* we hoped that we could resolve the question of the variability of x-rays stars once and for all," Giacconi said. "Also we could get a more accurate position, so that a visual counterpart could be found." If a visible star could be identified with Cygnus X-1, then the powerful optical telescopes could be brought into play; perhaps then, under scrutiny at two widely different wavelength bands, Cygnus X-1 would reveal its secret. *Uhuru* was pointed toward Cygnus X-1 on December 21, 1970, again on December 27, and again on January 4, 1971. These data showed that Cygnus X-1 was an erratically variable x-ray star. Sometimes it appeared to pulse rapidly and regularly with a certain period; another time it seemed that the pulse period had changed; on still other observations, there was no evidence of any periodicity at all.

They began to search the data on other sources for evidence of periodic variations. One of the sources they chose to study was in the constellation of Centaurus. This source, called Centaurus X-3, had been, like Cygnus X-1, the subject of conflicting reports in the literature. Everyone who had looked at it had gotten a different result for its x-ray intensity.

"I remember coming into work one morning and finding Ethan [Schreier] and Riccardo looking at the computer printout from the previous night's analysis," Harvey Tananbaum recalled. "The data showed that the intensity of Centaurus X-3 was varying in a very regular pattern. It was cycling on and off every five seconds or so."

Further observations over the next few months revealed that the period of pulsation was changing in a regular way with a two-day cycle, and that the source was turning on and off every two days. Suddenly it became clear what was happening. The x-ray star, presumably a neutron star, was in orbit around a large companion star. When it passed behind the large star, it disappeared from view, only to reappear a short time later. The regular changes in the period can be explained by the Doppler effect, according to which the frequency of a moving source depends on its motion with respect to the observer.

A familiar, though not so popular, use of the Doppler effect is in police radar. The radar equipment sends out waves of a certain frequency in the direction of a moving vehicle. When those waves reach the vehicle, they are reflected back to the radar. The vehicle becomes a moving source of reflected waves. The change in frequency of the transmitted and received radar waves is used to calculate the vehicle's speed.

Giacconi's group used the Doppler effect to interpret the regular changes in pulsation frequency of Centaurus X-3 and to show that it was in a very tight orbit around a companion star. This information was broadcast to the astronomical community in the form of an International Astronomical Union telegram, a service that keeps all major observatories in the world informed of new discoveries. Soon the companion star was identified. It was a massive blue star that showed evidence of orbital motion with a period of 2.1 days, the same as that of the x-ray star. With this information, the mass of the neutron star could be estimated: it was approximately the mass of the sun, that is, the expected mass of a neutron star.

During the next few months Giacconi and his colleagues discovered another binary x-ray pulsar with similar characteristics. The presence of a very nearby companion star and the absence of a surrounding nebula, as well as other characteristics, convinced astronomers that the binary x-ray pulsars were not working in exactly the same manner as the Crab Nebula pulsar and the other radio pulsars. In the model that is now generally accepted, the x-ray stars are neutron stars whose energy is produced not from their own internal heat but from infalling matter from a companion star. When a blob of matter falls through the intense gravitational field surrounding a neutron star, it acquires tremendous energy, owing to the acceleration of gravity. For example, if a blob of matter having the mass of a marshmallow (about 30 grams) were to fall onto the surface of a neutron star from a great distance, it would release the energy equivalent of a small atomic bomb. This energy would be converted to heat and would produce an intense hot spot on the surface of the neutron star. This hot spot would glow in x-rays.

The matter that produces the hot spot comes from the companion star. In a few cases, the two stars are so close together that they are practically touching. The matter on the surface of the companion star is torn between the gravitational force of the companion and that of the neutron star. This competition can distort the shape of the companion star until it looks more like an egg than a sphere. The same effect occurs, though on a much smaller scale, in the earth–moon system. The gravitational force of the moon pulls the earth slightly out of round, by about 30 centimeters. This distortion shows up as the high and low tides of the ocean. For two stars almost in contact, the tides can be enormous, and matter can be pulled away from the companion star by the neutron star. This matter usually forms a disk around the neutron star and, guided by the magnetic field of the neutron star, is funneled down onto its surface, where it produces an intense hot region that glows in x-rays. The rotation of the neutron star causes this spot to appear and disappear periodically, producing the regular x-ray pulses that are observed.

Did this model also explain Cygnus X-1? Not exactly. Groups from MIT and Goddard Space Flight Center had made rocket flights with detectors designed especially for the purpose of detecting rapid, regular pulsations from Cygnus X-1. They had found pulsations, but they showed no regularity. A companion star was found, but studies of this star's movement showed that the probable mass of the x-ray star was ten times that of the sun. A neutron star cannot possibly have such a large mass; it would immediately collapse to form a black hole. The observations showed, then, that the x-radiation from Cygnus X-1 could not possibly be produced by the infall of matter onto the surface of a neutron star. It could, however, be produced by matter falling into a black hole. The rapid irregular pulsations could be caused by hot blobs of matter swirling around the black hole before they disappeared beyond its gravitational horizon.

Cygnus X-1 thus provides very strong evidence for the existence of a new, bizarre type of object. Robert Oppenheimer and Harlan Snyder had shown theoretically in 1939 that a collapsing star of sufficiently large mass "tends to close itself off from any communication with a distant observer," and in the 1970s John Wheeler coined the name "black hole" for such objects. Except for a few scientists, such as Kip Thorne, Wheeler, Remo Ruffini, and especially Yakov Zeldovich and Igor Novikov, most scientists remained skeptical of the existence of black holes. They believed that as-yet-unknown effects would intervene to prevent the formation of such objects. But the combined observations from radio telescopes, optical telescopes, and the *Uhuru* x-ray satellite quickly dissipated this skepticism, and most astronomers accepted the observations as compelling evidence for black holes.

Scientists also generally agreed that the enigma of the x-ray stars had been explained. These objects are members of binary star systems in which matter streams from a normal star onto the nearby x-ray star, which is extremely dense owing to its collapse. In most x-ray star systems, the collapsed star is a neutron star, though in rare instances it is a black hole.

The two-year period after the launch of *Uhuru*, when the mystery of the x-ray stars was solved and the first black hole was discovered, was, in Giacconi's words, "the scientific highlight of my career. It was the most mystical moment, when we suddenly understood."

_10

If the *Uhuru* successes were the most mystical moment of Giacconi's career, they also saved the cause of the project toward which he had been pointing his whole career. "The x-ray telescope was what made it psychologically possible to work so long and hard in the 1960s," he said.

But the x-ray telescope observatory came perilously close to slipping away from him when the High Energy Astronomy Observatory program was abruptly canceled on January 2, 1973. How close is impossible to say. Given Giacconi's energy and drive to succeed, one cannot help but feel that somehow he would have saved it. On the other hand, the length of time it takes to get a large project going within NASA is seemingly impossible to overestimate. The scientific community must be convinced, support within NASA must be marshalled, and Congress must be persuaded to approve the funds. The whole process can take ten or fifteen years. If the cancellation had stuck, then Giacconi would have had to start the process anew— perhaps not at the beginning, but close enough to deal a crushing blow to the morale of his tightly knit group.

The decision to cancel the High Energy Observatory project was made for a combination of reasons relating to one fundamental issue— money. The Nixon Administration had put a ceiling on executive-agency spending for the fiscal year 1974. This ceiling was less than the amount of money appropriated by Congress, so many agencies, including NASA, found themselves short of funds. In NASA's case, the shortage was exacerbated by staggering cost overruns on the *Viking* program. There was no question in the minds of NASA administrators that this scientifically important, high-visibility effort to land a spacecraft on Mars in the bicentennial year of 1976 had to succeed. Nor was there any question of going to Congress to get an increased appropriation. The extra money would have to come from within NASA. The only program with the kind of funds that would be needed to rescue *Viking* was the High Energy Astronomy Observatory, which was itself getting more expensive every day. If it were canceled, NASA administrators reasoned, then two problems would be solved in one stroke: the *Viking* program would be rescued, and another expensive program would be eliminated. On January

2, without warning to NASA officials or scientists outside NASA who were involved in the program, the High Energy Astronomy Observatory program was canceled.

Dick Halpern's reaction was visceral. "After living and breathing HEAO for three years, the shock was too much. I went to the men's room and threw up." Jesse Mitchell's reaction was that of a veteran of the sometimes internecine NASA politics: never take the first word of cancellation as the last word, or in the words of Yogi Berra, "The game isn't over till it's over." With Halpern's help, he sought to convince NASA to change the status of HEAO from "canceled" to "suspended."

Mitchell, who has been described as "a kind of guy who would play all kinds of games within the system," moved quickly and decisively. He pleaded his case before the upper management of NASA. He pointed to the stunning discoveries by the *Uhuru* x-ray satellite. He showed them copies of a strongly worded resolution from the American Astronomical Society that reaffirmed the broad support for HEAO among astronomers and protested the cancellation. He produced plans for a scaled down version which he said would cost only one third as much as the original program, yet would do almost as much science and would satisfy the astronomers. And then he played his final trump: he threatened to resign if they did not give the program an eighteen-month stay of execution.

On February 20, Mitchell got his answer. They would give him eighteen months to come up with a detailed, workable alternative to the original program, at one-half the cost. He had won the first round. Now on to the second, in which he would have to persuade the scientists to accept the drastically reduced program peaceably. The history of HEAO gave him his outline for action. From the beginning, there had been three major spokesmen: Frank McDonald from Goddard Space Flight Center, who wanted a large cosmic ray experiment, Herb Friedman from the Naval Research Laboratory, who wanted a large, super-*Uhuru*-type x-ray detector, and Giacconi, who wanted an x-ray telescope. McDonald was out. His cosmic ray experiment was simply too heavy to be accommodated by the smaller rockets that would have to be used. Besides, he was a highly respected NASA science administrator, a team player who could be counted on not to cause trouble. Friedman and Giacconi were another matter. They were both independent, articulate, and skilled politicians, as well as respected scientists and fierce competitors. Somehow Mitchell had to make them both happy; either one of them could and probably would create such an uproar as to cause NASA to back off from the whole program.

"By that point, I was prepared to battle to the end," Giacconi said, "to say, let's fly the telescope first, let's not have any other mission. Instead, what happened was that they [Mitchell, Halpern, and Schardt] got Friedman and me in this smoke-filled room and kind of—you have to understand Mitchell—he was looking for a kind of negotiated deal. He wanted to make sure that the two major exponents of the field would have a deal which they could both live with. Basically, Friedman wanted to fly his detector, right? So I said, 'I think that's an excellent idea.' And then Friedman said, 'I think we ought to fly the telescope mission, too.' And I said," Giacconi laughed at the memory of the backroom meeting, " 'I think that's an excellent idea.' "

It was decided that Friedman's experiment would go first, since it was technologically simpler and, as a survey mission, was a natural prelude to a telescope mission that would make detailed studies of selected objects, some of which might be discovered in the survey. "I had been concerned all along about being third in line," Giacconi said. "It was too far down the line, too much could happen. Now we had a chance to move up a slot. Given what had been going on and given that no other x-ray proposals were in the offing, I felt that it was a satisfactory compromise."

"What about the rest?" Giacconi and Friedman asked.

"That will go as it will go," Mitchell said. Meaning, Giacconi said, "Let the dust settle where it will."

A few weeks later the principal investigators on the old HEAO missions were called to a meeting at NASA to hear how it would go. Walter Lewin of MIT was there: "The meeting had the atmosphere of a funeral," he said. "No one knew for sure who was going to be on, and who would lose his head."

George Clark, the principal investigator from MIT in the AS&E-Harvard-Smithsonian-Columbia-Goddard-MIT consortium, felt that not everyone was in the dark. Perhaps it was because Giacconi and Friedman seemed a little calmer than the others, or perhaps it was something else. Whatever the reason, Clark said, "I always felt that Riccardo and Herb knew more about what was going to happen at that meeting than the rest of us."

Mitchell then gave the speech that has become memorable among x-ray astronomers. "It was almost unbelievable," Lewin remembered. "He started off by saying that he had talked to his wife the night before, and he had said, 'Gee, what am I going to tell these scientists? They are all strong and good men, and here I have to tell them that certain things can't happen.' Then he started talking about a tree and manure, and that there was smell from the manure, yet something good was happening. I couldn't believe it."

Then Mitchell switched metaphors, this time to a sinking ship. "He said," Giacconi remembered, " 'You know guys, the ship is sinking. We are going to the lifeboats, and some are going to be left out of the lifeboats, and what we are telling you is that we are asking even those who are left out of the lifeboats to swim a little and help push.' " Giacconi laughed and shook his head. "That was what he said."

Then Mitchell told the scientists which of them were in the lifeboats and which of them were left out. One scientist lost two experiments and with them a small empire. He put his head in his hands and wept. Lewin was also among those left out. He and other scientists began to question Mitchell sharply about their methods for arriving at their decisions. The discussion became heated and the meeting soon dissolved into chaos. A break was called. All it accomplished was to allow the commotion to spill out into the hallways, where the scientists literally pushed and shoved to get the ear of a NASA official so they could plead their case. When the dust finally settled, several months later, a new HEAO program had taken shape, and Lewin and several others had managed to scramble back on board.

The first mission, *HEAO-1*, would be a scanning mission that would survey and map X-ray sources throughout the sky over a wide range of x-ray energies. Scheduled to fly in 1977, it included four cosmic x-ray instruments: Friedman's large-area x-ray detector; an experiment sensitive to a wide range of x-ray energies, with Elihu Boldt of Goddard Space Flight Center and Gordon Garmire of the California Institute of Technology as principal investigators; an experiment to determine the position and structure of sources, with Herbert Gursky of Harvard–Smithsonian and Hale Bradt of MIT as principal investigators; and a high-energy x-ray detector, with Laurence Peterson of the University of California, San Diego, and Lewin as principal investigators. Frank McDonald was chosen to be the project scientist; in this capacity he would act as a liaison between the scientists and NASA headquarters.

The second mission, *HEAO-2*, would be the focusing x-ray telescope proposed by Giacconi's consortium. It was scheduled to fly in 1978. The third and final mission, *HEAO-3*, scheduled to fly in 1979, would carry three separate experiments designed to study gamma radiation and cosmic rays. In July 1974, a year and a half after word came down of the cancellation, major funding began again.

While the battle to save the HEAO program and with it the x-ray telescope was being fought and won, Giacconi was fighting another battle, with less success. Over the course of twelve years at AS&E, he had built the Space Science Division from four people working in a converted milk-truck garage to an internationally acclaimed x-ray astronomy group of seventy or eighty people that occupied four floors in a modern, high-rise office building in downtown Cambridge, as well as laboratory space in suburban Woburn. It was a cohesive group molded to a large extent in Giacconi's image: hard-driving, dedicated, efficient, and able to deliver the goods. Many scientists in academia, some of them in positions of influence, expressed doubts as to whether it was "proper" to "do science for profit." Despite their misgivings, AS&E's share of NASA money for research in x-ray astronomy continued to grow, because they produced results. The consequence was admiration, tempered with more than a little fear and envy.

This attitude was enhanced by stories that came back from some of those who visited AS&E or worked there temporarily. In the glory days of *Uhuru*, new scientists were hired and visiting scientists from all over the world passed through AS&E. Some found the environment stimulating. Paul Gorenstein, one of Giacconi's colleagues, recalled, "He [Giacconi] motivated people, got them to do things. Whether he did it by inspiration or example, he got a lot of people to deliver their best." Other newcomers and visitors were shocked at the rough and tumble, no-holds-barred atmosphere as the group discussed the meaning of the data that was flooding in and the priorities for new observations.

"We created what we called the *Uhuru* spirit," Giacconi said. "With the *Uhuru* spirit you absolutely let the other guy have it, but it was done in a spirit of searching for the truth, and of trust and love and help. People tended to see us as a tightly controlled monolith, but within the group there was always a lot of freedom. And most of all, it was a rational world. The validity of your ideas was what counted, not personalities or company politics."

Because Giacconi had risen to the position of executive vice president and was a member of the board of directors, and because the Space Science Division was responsible for most of the company's income, he was able to effectively shield the scientific staff from the politics of the company until 1973. After that, conditions worsened.

"It became impossible. I decided that I wanted to move. I talked to Leo Goldberg [director of the Harvard College Observatory]. He suggested that I should apply for the directorship of the Smithsonian Astrophysical Observatory [which shares a sprawling complex on Observatory Hill in Cambridge with the Harvard College Observatory and the Harvard University Astronomy Department], since Fred Whipple was retiring. I talked to Dean Dunlop, then the dean of Arts and Sciences at Harvard, about it and he was agreeable. In the meantime, Goldberg decided to retire, and the idea of unifying HCO and SAO and the Harvard Astronomy Department into a single Center for Astrophysics with one director became popular. George Field was chosen as the director, so I became an associate director."

"Another factor in my decision to leave AS&E was that the corporation was turning more and more toward commercial enterprises. It was planned that *Einstein* [the name given the x-ray telescope observatory after launch] would be a national facility and I believed that this could be more easily accomplished in an institutional setting such as the SAO. I also believed that the scientific staff would benefit from the better intellectual milieu of Harvard, with graduate students, seminars, etc. Now, in retrospect," he added mischievously, "it's arguable whether Harvard was in fact a better intellectual milieu, but that's what I thought anyway."

The feelings were not exactly mutual. Certainly, those in high positions at Harvard–Smithsonian were convinced that the addition of Giacconi and his group would renew the vigor of the once great institution that had over the years grown complacent and gradually slipped from the forefront of astronomy. But many of the staffers viewed Giacconi's coming as an invasion. He secured tenured appointments for four of his senior scientists, and almost all the remaining scientists chose to come with him in positions paid from

NASA contracts that would be transferred from AS&E to Harvard–Smithsonian. This group became the High Energy Astrophysics Division of the Center for Astrophysics. Because of the size of the contracts held by Giacconi's group, and because of the scientific productivity of the group, the High Energy Astrophysics Division immediately became, by any objective standard, the pre-eminent group at the Center—but not the most popular group.

Giacconi and his colleagues, who had sought refuge in the intellectual milieu of Harvard–Smithsonian from corporate bloodletting, soon found themselves beset by academicians who interpreted the repainting of the walls as proof of Phillistine hubris. They decried the lack of individuality in the group and dismissed the work of Giacconi's colleagues as "mere engineering" or "the intellectual equivalent of weapons research." Giacconi responded in kind. And so it went. Over the years, tensions eased; but today, more than ten years later and several years after Giacconi's departure to become the director of the Space Telescope Science Institute, remnants still exist.

The opinion that Giacconi's management style suppressed individual initiative is an outsider's view that is not shared by those closest to him. His co-workers appreciated his tolerance of different styles and approaches to scientific research. "He is demanding of himself and of the people who work for him," Harvey Tananbaum told us, "but at the same time he is very, very supportive. He gives you the opportunity to develop your own scientific, intellectual, managerial, and technical skills, all of which are required for large projects . . . People who worked with him right down to the junior level were encouraged to be creative . . . It was not a directive process . . . I think we all had tremendous trust and confidence in Riccardo, in his judgment as a scientist and a leader, and also as a human being, that he wouldn't do things to knowingly hurt us, and was interested in our growth as individuals."

In the meantime, Giacconi's group, working with the other members of the x-ray consortium, had an x-ray telescope to build and four years in which to do it after the funding resumed in July of 1974. In spite of the cutbacks involved in restructuring the HEAO program, Giacconi and his colleagues still hoped to produce an observatory that would be the x-ray equivalent of the best radio and optical telescopes. To do this, they would have to build a complex of sensitive x-ray detectors; it was not possible to build a single detector that could make fine x-ray images and detailed spectral measurements, see a large field of view, and detect very weak sources.

The x-ray consortium decided on four primary instruments. One was a high-resolution imager (HRI) that could make detailed x-ray images with a resolution comparable to that of a large optical telescope. The second was a detector that could make broad-brush images of x-ray sources at more than one x-ray energy and had the sensitivity to detect very weak sources. The Harvard–Smithsonian group had responsibility for these two instruments. A third could, for strong sources, study the fine details of x-ray spectra. The MIT group had responsibility for this instrument. Finally, the solid-state spectrometer could make detailed spectral measurements over a broad range of x-ray energies. The Goddard Space Flight Center group had responsibility for this instrument. Incoming x-rays would be focused onto one or the other of these detectors by an x-ray mirror assembly of four nested pairs of highly polished cylinders, each about two feet in diameter. The Harvard–Smithsonian group had responsibility for assembling the x-ray mirror.

Though different members had responsibility for different instruments, it was agreed that all scientists in the consortium would share in the data from all the instruments. This arrangement, which sounds eminently reasonable, represented a radical departure from the usual practice. In the formative years of x-ray astronomy, the different teams had been keen competitors, both for scientific rewards and for research funds, and had joined together only on rare occasions. This tradition had continued into the satellite era: *Uhuru* and the small x-ray satellites that followed it either were one-group satellites or carried instruments from several independent groups. In these cases, which included *HEAO-1*, the groups merely shared space on the satellite, like the owners of apartments in a condominium; each group built its own instruments and each had exclusive rights to the data from those instruments.

Giacconi realized that the condominium approach would not work for the x-ray observatory. The groups would have to accept a marriage of convenience and live together like a family. The entire complex of instruments would have to be integrated into a complete whole with exquisite care. As technological and monetary crises arose, differences would have to be aired in open, frank discussions, and compromises would have to be made that would satisfy everyone. Frank McDonald, now NASA's chief scientist, praised Giacconi's work in this area. "Riccardo did an outstanding job getting everyone to work together." If one member of the consortium had a problem, they were supported by the other members. When the instrument proposed by the Columbia group became a victim of the reduced budget, leaving them without an instrument on the observatory, they nevertheless retained their rights to a share of the data as members of the consortium.

As in any family, life wasn't all blue skies and sunshine. Steve Holt's experiences were typical. Holt, who played a double role on HEAO as a member of the Goddard part of the consortium and as the NASA project scientist for *HEAO-2*, often found the negotiations within the consortium exhausting. "Everyone was extremely capable of defending his own turf. I learned to drink martinis as a result of those meetings. I would be so drained after the meetings in Cambridge that I would try martinis on the plane. For about a year, the only time I would drink was on the plane." Several years later, Holt, though he still likes an occasional martini, remembers the HEAO experience fondly. "On the whole, it was a lot of fun, from beginning to end."

Most of the instruments planned for the x-ray telescope observatory were adaptations of devices that had operated in other applications. Not so the high-resolution imager. It had no prototype.

"We started with a vidicon system," Giacconi said. "That was the way we sold the project." This system, which was basically a sophisticated television camera, was used to make simulated x-ray pictures of galaxies in the consortium's presentation before NASA in 1970. After a while, though, "it became clear that it would be almost impossible to get that system working in orbit. We thought about it for a little while, but it would've been almost impossible. Then we read about this wire-spaced readout system." This system, which had been developed by a group of scientists from the University of Leicester in England, used a microchannel plate array. A microchannel plate array is a collection of millions of tiny glass tubes that are coated with a material that converts x-rays into electrons. When an incoming x-ray enters the tube and strikes this material, it kicks an electron free from the material. This electron continues down the tube, bouncing off the wall and producing an avalanche of millions of electrons. This avalanche emerges from the end of the tube as a charged cloud that spreads across a vacuum gap until it encounters a grid of wires strung like guitar strings below the microchannel plates. When the charged cloud encounters these wires, it produces an electric signal. The British group had shown that by analyzing the signal from the wires, one could figure out the line along which the x-ray hit the microchannel plate. From this a one-dimensional image of a source can be reconstructed.

"It occurred to us that we could make them two dimensional," Giacconi said. To do this they would have to construct a crossed grid of wires. "Leon [Van Speybroeck] started off a program in which he was trying to get industry to build very closely spaced, etched, so to speak, wire and empty space things. He went at it for six or nine months. And it was a failure. The industry just couldn't make it. So then, there was the fundamental turning point in which Steve [Murray] and I got together and we said, 'Why is it that we are trying to get these wires so closely spaced? Why couldn't we determine charge centroids with wires more widely spaced?' "

This breakthrough was possible because, in x-ray astronomy, the photons from a cosmic source arrive at a rate ranging from a few per second to about a thousand per second. By comparison, optical photons from a strong source can come in at a rate of millions per second. The low rate of x-ray photons meant that the charged cloud from each photon could be detected individually. And they did not have to pinpoint precisely where it hit on a fine grid. They could let the cloud spread out a little more, hit several points on the grid, find the centroid of these points with a computer, and reconstruct a two-dimensional image of the source. This could be done with wires a few tenths of a millimeter apart.

"Then we wouldn't have to go to industry to get closely spaced things. We could do it in our lab. And that's the way it was done." In Murray's words, "We just kept trying things until something worked. We really played the basement inventors. It took a few years, but after going down many blind alleys, finally we succeeded. In the end it turned out that the simplest ways were the ways that worked." For example, the detector required a grid of wires that had to be wound 128 wires to the inch, evenly spaced. How to space them evenly? Wind a double strand of wire and then unwind one strand.

The heart of the observatory was the mirror assembly. The complement of detectors had been designed so that if one of them failed, the mission would not be a disaster. But if the x-ray mirrors did not work, the mission would be a failure. The scientist primarily responsible for seeing that the mirror assembly did work was Leon Van Speybroeck. Van Speybroeck had been working on x-ray mirrors since 1967, when he joined the AS&E solar x-ray group. In the years since then he had worked with Giacconi, Giuseppe Vaiana, and others to develop the x-ray mirror assembly that had worked so well on *Skylab,* and he had come to be recognized as an experimental genius in the esoteric practice of designing x-ray mirrors. The final design involved four concentric sets of nearly cylindrical surfaces, each about two feet in diameter. The interior surfaces of the mirrors were to be the shape of a parabola followed by a hyperbola. The curves would be so subtle that the mirrors would resemble cylinders with a slightly conical internal surface. X-rays entering the telescope would encounter the mirror at a grazing angle and would reflect first off the parabolic portion of the mirror and then off the hyperbolic portion. These two reflections would focus the x-rays to a point. Four such mirrors would be nested inside one another, like successively smaller measuring cups. For the x-ray telescope observatory, Van Speybroeck's job was not merely to design the best x-ray mirror ever made; he had to do it for the lowest possible cost. And it had to be done on a rigid schedule. NASA officials, always fearful that the whole program might be cancelled again if the project went too far over budget or fell too far behind schedule, kept the pressure on.

"HEAO was always a problem because it was always a little underfunded," recalled Albert Opp, NASA program scientist for the HEAOs. "As technical problems came along, they were very hard to cope with. The hardest problems were the *HEAO-2* overruns. It was a bigger and more difficult mission than everyone had estimated. The cause of the overruns was the optical system itself. It was an exceedingly sophisticated, exceedingly complicated undertaking."

"We had to solve a hundred thousand technical problems," recalled Harvey Tananbaum, the scientific program manager for Giacconi's Harvard–Smithsonian group. There was the problem of finding a manufacturer who could make large, high-quality glass cylinders with approximately the desired shape. Eventually the West German optics company Heraeus Schott Quarzschmelze came up with an intricate process that worked. These cylinders were shipped to Perkin-

Elmer Corporation in the United States to be precision-ground to conform to the peculiar subtle curves needed to focus x-rays, and polished so that no imperfections over one ten millionth of an inch remained. One of the cost-saving cuts in the program was that there could be no test mirror—this one group of mirrors was it. The result was a double-barreled anxiety attack among the scientists, especially the cautious, meticulous Van Speybroeck. First of all, they would not have the 'learning experience' with the test mirror to give them confidence before the final polishing and testing of the flight mirrors. Second, there would be no backups in case something happened to the mirrors. "Leon hovered over those mirrors like a mother hen," Giacconi remembered with a smile. "He literally slept with them," Tananbaum recalled. "He never let them out of his sight during the crucial periods." Van Speybroeck, for his part, acknowledged that "it had to be a burden upon the Perkin-Elmer staff to operate with such constant consumer presence." Yet he felt that "if I were doing it over, I would want to control certain environments even better." When NASA complained that Van Speybroeck was being too much of a perfectionist, he argued that it would be cheaper in the long run to do it right. Still fearful that the coating and assembly phases would take too long and cause the project to fall further behind schedule, NASA ordered the polishing to be stopped one day early, a decision that caused the x-ray mirrors to be less efficient at high x-ray energies than they otherwise would have been.

We asked Giacconi if there was ever a time of panic, a time when he thought they were not going to be able to pull it off. "There were periods of nervousness," he said, "but never panic. There was the issue as to whether the telescope would work in a space environment. We really didn't know until we put it into the chamber at Huntsville."

It was at Marshall Space Flight Center in Huntsville, Alabama, late in the summer of 1977, that the entire observatory would be tested under conditions similar to what it might encounter in space. The scientists would be especially watchful for the types of problems that plague space experiments: gas leaks from any of the counters that contained gas confined by thin, fragile windows; and "corona," or high-voltage arc discharges from one electrical component to another. In a high vacuum, corona are a persistent danger. According to Steve Murray, "Corona are the most insidious problem in space applications; they are probably the major cause of in-flight failures."

The test facility at Marshall, which had been built there especially for testing the x-ray observatory, consisted of a huge vacuum chamber 20 feet in diameter and 40 feet long where the telescope would be housed. Connected to this chamber was a 1,000-foot-long pipe, at the end of which was a source of x-rays. The long pipe was necessary to allow effective focusing of the x-rays. By placing an x-ray source of known intensity and spectrum at a known angle in front of the telescope, they could measure the response of the telescope and compare it with theoretical calculations. The consortium planned to make more than a thousand such measurements to thoroughly test the telescope over a wide variety of conditions. They scheduled six months in which to make these tests.

Once again the pressure of time intervened. The project was now six months behind schedule. And time equals money. NASA shortened the test time to one month. "Impossible!" Giacconi protested. If they worked double shifts and wasted absolutely no time, it would take 31 days. "Thirty days, no more," came the reply from NASA. Not wanting to short-change the testing procedure, Giacconi asked Van Speybroeck to develop a computer program to see just how quickly all the tests could be performed under optimum conditions. The answer: they could do 1,397 tests, with an average time of 18 minutes and 30 seconds each, in 18 days, if they worked around the clock in overlapping 13-hour shifts. This would allow 12 days for solving problems that would inevitably arise. Fred Speer, the HEAO program manager at Marshall who had often clashed with Giacconi during the development of the observatory, agreed to help, and provided staff support to keep the calibration facility running 24 hours a day. By the end of the 30-day period, the exhausted scientists and engineers felt confident that they had an observatory that would work.

One valuable by-product of the crash effort to test the observatory was the development of an efficient data-handling system. The scientists knew that the testing would generate an avalanche of data and that these data would have to be analyzed immediately to see whether follow up tests would be needed. Throughout his career, Giacconi had recognized the importance of efficient data-handling systems in space astronomy. He had assigned high priority to the development of software to handle the data from *Uhuru;* this effort had paid off handsomely in the investigations that had led to the discovery of binary x-ray stars and a black hole. Their ability to process the data almost immediately and respond rapidly to surprising observations led to a quick solution of these problems. In the same way, he determined at the beginning of the x-ray observatory project that a large part of the effort would be devoted to the development of the data-handling system, so that they could process the data and produce meaningful scientific results immediately after launch. The upcoming testing period would also provide a crucial test of the computer software that had been developed to process the data. He assigned more scientists to the software effort, with the mandate of having the system ready by the time testing began. The software group met the challenge. By the end of the testing period, the consortium not only knew the observatory would work, but they knew that they had an efficient, thoroughly debugged data-handling system that would allow the scientists to go smoothly from observations to data analysis.

The x-ray observatory stands in marked contrast to the Very Large Array (VLA), where the shortcomings of the data-handling system led to frustrating bottlenecks in the data pipeline. In fairness to the VLA designers, the differences in the two projects should be made clear. The x-ray observatory was a complex of instruments that was scheduled to fly for a limited time, so it was imperative to get the most possible information out of every minute the observatory was in orbit. Moreover, once in orbit, the instruments could not be modified. The VLA, on the other hand, is an open-ended, ground-based project. There was no particular urgency, other than the urgency to use the instrument, to get the data-handling system perfected ahead of time. More importantly, the capability of the instrument, as well as computer capability and computer techniques, would undoubtedly be upgraded as time went on; so a large effort to develop the ultimate data-handling system would have been expensive and ultimately would have become outdated anyway. While these characteristics of the VLA perhaps explain why its data-handling system has been developed so slowly, we are convinced that future planners of ground-based observatories will have to devote considerable effort to plans for handling the flood of data that new leaps in technological capability inevitably bring.

_11

Successful practitioners of the art of persuasion, be they salesmen, evangelists, politicians, or managers of large-scale science projects, understand the power of symbols, labels, and names. Shakespeare's Juliet protested against this power, arguing that "That which we call a rose, by any other name would smell as sweet." But Juliet knew that most people put great stock in names. In times past, names gave fairly specific information about parentage, place of origin, and vocation. We still put a great deal of thought into naming our children, often picking the name of a saint or some other revered person, such as our parents or ourselves. The American Indians had the somewhat more sensible tradition of allowing individuals to choose their own name as they entered adulthood.

NASA's record on names has been mixed. On the one hand, there have been the *Mercury, Apollo, Pioneer, Viking,* and *Voyager* programs—names that evoke the spirit of adventure and exploration. On the other hand, there have been the acronym programs—*OSO* (Orbiting Solar Observatory) *1, 2, 3, 4,* and so on, *SAS* (Small Astronomy Satellite) *1, 2, 3*—whose names evoke images of bureaucratic blandness. The names for the High Energy Astronomy Observatories were in the latter tradition: *HEAO-1, HEAO-2, HEAO-3.*

Giacconi had flouted this tradition with *Uhuru.* The official name for the first x-ray satellite was *SAS-1,* but he felt that this unique, trail-blazing instrument deserved a unique name. NASA officially refused to allow the satellite to be renamed, and insisted that Giacconi refer to *Uhuru* as a nickname. Giacconi refused to refer to it by any name *other* than *Uhuru.* The NASA bureaucracy eventually came around (many NASA scientists and managers had approved of the name *Uhuru* from the beginning) and now routinely refer to the satellite as *Uhuru* in official publications. Likewise, Giacconi knew that the x-ray observatory would be a unique facility, and he wanted it known by some name that would symbolize the historic contribution it would make to our knowledge of the universe.

"Harvey Tananbaum suggested *Pequod,* Captain Ahab's ship in *Moby Dick,*" he remembered. "I liked it. I liked the images it evoked—

American Indians, New England whalers, a blending of cultures, a sense of adventure." Giacconi was also well aware of other images evoked by the name.

"It was really funny," a fellow astronomer commented. "He thought *Pequod* represented Massachusetts, the American Indian, and so forth. *Pequod* to me represented a ship captained by an egomaniac."

"Of course I was aware of the inevitable comparisons between myself and Captain Ahab," Giacconi said, "but I didn't mind. After all, what did Ahab do? In the tradition of Ulysses in Dante, he imperiled his own crew in search of knowledge. Anyway, I called Noel Hinners [then associate administrator for space science at NASA]. He said it would take two years to get the name changed, and besides, NASA was not particularly enthusiastic about associating one of its satellites with a white whale." An attempt to get President Carter involved in naming the satellite also came to naught, but Giacconi was determined to give it a name, albeit an unofficial one. "I knew that if the name was going to stick, it would have to be one that everyone in the consortium liked, so that they would use it in their publications. So we went by the democratic process. We asked the members of the consortium to submit names; we prepared a list and took a vote. By a large margin, the name chosen was *Einstein*, in commemoration of the centennial of his birth, which would occur a few months after the launch.

The launch of *Einstein* was scheduled for a few minutes after midnight on Monday, November 13, 1978, from Kennedy Space Flight Center, Cape Canaveral, Florida. The members of the consortium gathered there, many with their families, to watch the culmination of years of work.

"George [Clark] was there; so was Bruno"—Bruno Rossi, who nineteen years earlier had suggested to Giacconi that x-ray astronomy might be a field worth investigating and had co-authored with him the first paper describing the design and potential of an x-ray telescope. "George used the occasion to bring us back together again. He proposed a toast to the grandfather and the father of x-ray astronomy. It was a nice gesture. One of the most moving things, though, was the presence of my son, who had come down to see what his old man had been up to for all those years."

On the night of November 12, shortly before midnight, Giacconi took his position in the blockhouse near the launch pad. "During the countdown, I was given one of the 14 different channels used to follow the countdown. All the channels were open all the time. I heard lots of hair-raising things, things such as 'Leak in hydrogen line,' 'anomalous voltage level.' It was alarming, but there was nothing I could do. Pat Henry best knew about the instrument; so he was nearest to me in the control room. Leon wasn't there; he couldn't stand it. He had gotten so nervous during testing that he developed stomach problems and couldn't go to Marshall. He didn't want to go to launch, either. He stayed at Goddard, where he had a direct line to me in case he saw anything wrong on the monitors. I was terrified. I had the impression of total helplessness. There were all these guys, 500 or more in number, all of which could kill you. You hope everyone is well motivated. You are totally dependent on people you never heard of. The Safety Range Officer could blow it up."

At 12:22 a.m. on November 13, he heard the words, "Ignition."

Harvey Tananbaum was watching from the bleachers. "I felt a chill run through my body when I saw the tower light up at T-1 with ignition. It was like lightning and thunder. You couldn't hear anything for several seconds. By then the vehicle had already lifted off. It began to rise very, very slowly, gathering speed. It was a powerful visual, auditory experience. You momentarily forgot that it was carrying your life's work, as it were, under the nose cone. We got a report a few minutes later that the different stages had fired properly and that the orbit would be achieved. That was really a very satisfying moment."

The next critical moment would be "activation," when the instruments were turned on. Then and only then would they know for sure that they had an observatory that worked in space, some 200 miles above the earth. Activation is always an anxious, nail-biting moment, because if the spacecraft does not work, because of corona or some other problem, there is very little that can be done about it. The activation period for *Einstein* produced one nervous moment when a preliminary test indicated that the star trackers—cameras that track bright stars and provide crucial information as to exactly where in the sky the telescope is pointing—were not working. Fearing an electrical malfunction, Murray and Schreier, the two experts on observatory operations, ordered the star trackers turned off and began agonizing. Had they shorted out? Or was it a mistake in the software? Or some effect they had not considered? Finally, four days later, they realized that the cameras had been picking up the reflection of the moon off the Pacific Ocean during the checkout, and that they had been working properly. They were turned back on, and the telescope was pointed to the position of Cygnus X-1.

Giacconi was at Goddard, watching on a monitor. "I could see the stars streaking through the field of view," he recalled. Finally, eighteen years after he had dreamed of an x-ray telescope, he was seeing it in action. "It was an unbelievable sensation. A profoundly moving moment."

_12

The *Einstein* x-ray observatory was an immediate success. "The x-ray images from *HEAO-2* created a tremendous impact on the scientific community," said Albert Opp, astrophysics chief at NASA. "They will go down as milestones in the history of astronomy." Malcom Longair, director of the Royal Observatory in Edinburgh, Scotland, agreed, comparing the launch of *Einstein* with Galileo's first observations of the moon and the planets.

Over a period of two and a half years the *Einstein* observatory made more than 5,000 observations of objects ranging from comets in our solar system to quasars billions of light-years distant. On April 26, 1981, the gas supply in the thrusters used to control the orientation of the spacecraft was exhausted and the observatory ceased operations. It was, Giacconi felt, an unnecessary but calculated premature death. Other satellites, including *Uhuru*, had used magnetic torquing systems to orient the spacecraft. In these systems, magnetic fields are applied to bars in the spacecraft, causing it to turn. The power used to operate this system is derived from solar panels; so in principle it could last indefinitely. In practice, the only drawback with a magnetic torquing system is that it can turn the spacecraft only very slowly. This is no problem if everything goes as planned, but in an emergency it could prove fatal, because solar panels have to be pointed toward the sun every five hours, or the batteries on the spacecraft will run down. The magnetic torquing system represented an additional cost of about one half to one percent of the total cost of the msssion; it was deleted in the restructured program. Giacconi felt that, since the magnetic torquing system would prolong the lifetime of *Einstein*, a unique observatory which would not be duplicated for a decade or more, it was well worth an additional half a percent.

"I went in there and I said to them that the HEAO program was terribly important for research in x-ray astronomy. In particular, *HEAO-2* [*Einstein*] had to be made to survive as long as it could. And they said, 'Well here is another greedy scientist that wants more than what he can get.' All scientists look the same to them, and why should they believe me and not another and so forth?"

What about the advisory boards, such as the Space Science Board? "It never came to the Space Science Board," he said. "The Space Science Board has made a terrible mistake all the time, which is to think that it is enough to set down general guidelines, and that they will be intelligently followed and that they do not have to get involved."

But doesn't NASA have some committee available for consultation? "They don't want to ask them what to do. They want to be in charge. They want those guys out there to say, 'You should do this and that,' but they actively resist any—what should've happened, is that I should have gone to the Space Science Board. Please understand, I try to work within the system and ask them to be rational."

"I remember a conversation with Al Schardt. It went like this. I said, 'Look, we've got to prolong the life of *Einstein*. Suppose even a year or even two years, that would be fantastic. The scientific return would be enormous.' "

"And he would say, 'But then I would have to spend $11 million for operating the telescope each year. Don't you realize that I would have to spend another $11 million?' "

"And I said, 'But that's ridiculous, you've spent $100 million to put this thing up and it costs you only $11 million to keep it up an extra year!' "

"He said no. NASA does insist on running the mission as they see fit. The other point is that they were responding to the OMB [Office of Management and Budget], because OMB didn't want to have an open-ended mission. They wanted the mission to go up, do its thing, and finish. That's it. No more. Go away," he laughed.

"These problems are with us now. They cut down on the HEAO data-analysis money. Harvey [Tananbaum] fought very valiantly and supposedly got a million dollars restored. This year it fell in the crack and the million dollars got lost again, so he was once again fighting to get it restored. Meanwhile Hans Mark [NASA associate administrator for space science] goes to Congress. In response to some question, he says, 'Oh no, no. NASA knows how to run these programs. Look at the HEAO program. They went up, they got the data, and now it's all finished.' "

Giacconi laughed and shook his head. "So that attitude is still there. They have forgotten that *Uhuru* ran for four years and that we went on doing data analysis for maybe eight. IUE [the International Ultraviolet Explorer satellite] has gone on forever. The only one that is really coming out of the mold is the Space Telescope. Now, they are beginning to look at what it means to have permanent facilities in orbit. But back then, they weren't. NASA has only a few people in high places, such as Frank McDonald or John Naugle, who really understand about science and how you do it and the need for science. Most of the other guys have a very short-term kind of dedication to programs. They march to a different drummer. I mean, if at a certain point Nixon imposes a certain style which has to do with a tight budget, cut-the-cost discipline and so forth, then we get this thing that we can't have an open-ended mission for *Einstein* because we must be able to tell OMB a fixed cost. But that's insane, right?"

"In the case of *Einstein,* they selected a triple system failure. They selected low gas. They selected low orbit. I fought for a high orbit, but they said no. Finally, they put us on gyros which failed one after another—tock, tock, tock—so there was a triple system failure."

Despite these constraints, Giacconi and the *Einstein* team determined to extend the mission as long as possible. Schreier and Tananbaum, together with engineers at Marshall Space Flight Center, developed an ingenious system for minimizing the use of thruster gas. By carefully selecting the targets and scheduling observations, the build-up of the angular momentum or spin of the spacecraft could be greatly reduced, and the thruster gas conserved. This technique doubled the useful lifetime of the spacecraft, from one to more than two years.

We asked Giacconi about the scientific results of *Einstein.* What were the high points? "The results on the x-ray background was certainly one of them," he replied. The x-ray background is the reference level against which x-radiation from stars and galaxies must be observed. If we look at the night sky in visible light, we see a large number of bright spots plus a few diffuse patches of light on a black background. But when the sky is observed in x-rays, there is no equivalent of the black night sky. Instead, there is a bright diffuse emission, similar in its detailed properties to what you would expect from a 400-million-degree gas spread uniformly throughout the space between the galaxies. The uniformity of the background radiation means that it must originate far outside our galaxy. Since the discovery of the x-ray background radiation on the AS&E rocket flight in 1962 when the first x-ray star was discovered, astronomers have agreed that the x-ray background radiation contains vital clues to the origin and evolution of the universe. They have not agreed, however, as to what produces it. One theory is that it comes from hot gas spread uniformly throughout space. The total mass of this gas would be enormous; it would be very nearly enough to halt the present expansion of the universe and cause it to collapse billions of years from now. If this were so, the universe as we know it would be finite and would end in a cosmic fireball produced by the collapse.

An alternative theory is that the x-ray background radiation only *appears* to be a uniform glow. In reality it is produced by millions of distant sources of x-rays, such as quasars. The uniformity results from the sources' great distance—like the glow of a distant city. Giacconi had long maintained that an x-ray telescope would settle the controversy, because it would allow the individual sources to be resolved, if indeed they exist. One of the first goals of *Einstein* was to test this hypothesis by making long observations of blank fields of the sky, that is, regions empty of known cosmic sources of radio, optical, or x-radiation. If individual sources were producing the x-ray background, they should begin to show up as discrete pointlike sources of x-rays in the blank field surveys.

They did. The surveys analyzed so far indicate that at least 30 percent of the x-ray background, and possibly most of it, can be accounted for by quasars and similar objects. "I really believed that you had to image the x-ray background to study it," Giacconi said. "It was very satisfying to see that I was right."

"Then there was the satisfaction of having brought x-ray astronomy to a level of parity with other branches of astronomy. I heard a talk the other day by Marshall Cohen [a Cal Tech radio astronomer] about jets in quasars and active galaxies. It was amazing and gratifying to hear the use he made of x-ray observations to determine the size of the emitting region, and so forth. And in stellar astronomy, x-ray observations of stars are proving to be very useful."

Surveys using the *Einstein* observatory have shown that all stars are x-ray sources at some level. In contrast to neutron stars, which radiate most of their energy in the form of x-rays, most normal—that is, noncollapsed—stars radiate only a small fraction of their energy in x-rays. This radiation is believed to come from a corona (meaning "crown") of hot gas that overlays the visible surface of the star, the photosphere. The temperature of the photosphere of the sun is about 6,000° Celsius; the corona has a temperature of over a million degrees. Studies of the x-ray emission from the sun and other normal stars are therefore primarily studies of the coronas of these stars. These coronas provide a cosmic setting for investigating how hot gases are produced in nature and how magnetic fields interact with hot gases to produce flares, the spectacular explosions that release as much energy as a million hydrogen bombs. They also give us a means for probing beneath the surface of the star, because the heating of the corona of a star can usually be traced back to the interior of the star.

Since we cannot see the interiors of stars, we must deduce what is going on there from the radiation we observe from their surfaces and coronas. This is done by means of a model star, which is constructed according to some plausible theory; the predicted radiation from the star according to this model is then compared with the observed radiation. If the predictions match the observations, the model is deemed a success, and enters the books as the true description of how the star works. If the predictions do not match, the model is discarded and another model is sought. Of course, life is never so simple. The agreement between predictions and observations is never so good that astronomers can be confident they have fully explained the interior of a star. Besides, new observations are continually being made which put the models to newer and even more severe tests. Models, unlike champion prizefighters, cannot retire undefeated; they must meet every challenge or they will be abandoned. The x-ray observations of stellar coronas cover stars of many different types: stars such as the sun, stars much larger and much younger than the sun, and stars much smaller and much older than the sun. These observations have presented new challenges to the model builders. In the years to come, some of the old models will fail, and new ones will be brought forward. In the process, our understanding of the interiors of stars will deepen.

Another area where Giacconi feels that *Einstein* has made a creative contribution is in research on clusters of galaxies. Galaxies are not distributed uniformly through space, but clump together in groups and clusters. Our Milky Way galaxy and the Andromeda galaxy are part of the Local Group, a collection of 20 or more galaxies of which Andromeda and the Milky Way are the largest. About 60 million light-years away in the direction of the constellation of Virgo is the Virgo cluster of galaxies, which contains about 10,000 galaxies packed into a region only a few million light-years in diameter. About 10 percent of all galaxies in the universe appear to be members of rich clusters like this one.

One of the important discoveries made with the *Uhuru* satellite was that rich clusters of galaxies are pervaded by a tenuous gas that has been heated to temperatures up to 100 million degrees. It was not until the flight of *Einstein*, though, that it became possible to study the distribution of this hot gas in detail. The mass of this gas is comparable to the mass of the visible parts of galaxies, so it is a major dynamical component of a cluster. More significantly, we know that the gas must be confined by the gravitational field of the cluster, otherwise it would have evaporated away long ago.

This poses a major puzzle. The combined gravitational pull of the matter observed in the clusters using radio, optical, and x-ray telescopes is not sufficient to trap the hot gas in the clusters. Yet it is clearly trapped. The gravitational field is apparently 10 times stronger than one would predict on the basis of the matter observed to be present. Either the theory used to calculate the gravitational mass is wrong, or 99 percent of the mass in the cluster is in some dark, so far undetected, form. X-ray, radio, and optical observations have shown that this "dark-matter mystery" exists not just for clusters of galaxies but for virtually every galaxy in the universe. Not surprisingly, it is at present one of the hottest areas of research in astronomy today.

Astronomers also hope to gain insight into how clusters of galaxies are formed by studying this hot gas. Did individual galaxies form first and then come together in clusters, or did galaxies condense out of a large, pre-existing, cluster-sized matrix? This question is far from trivial; it affects our understanding of how the structure we observe today evolved from the hot cosmic soup of the very early universe. Some theories say that galaxies formed first; others give clusters priority.

Einstein made x-ray images of scores of clusters of galaxies. These images showed that galaxy clusters range in structure from irregular, clumpy groups to smooth, round spheres. These differences appear to be due to the evolution of the clusters, and strongly suggest that the galaxies formed first; later, they clumped together into the clusters we observe today. In irregular clusters, such as the Virgo cluster, this clumping process has only recently begun, whereas in Coma gravitational forces have pulled the galaxies into a uniform, round shape. In between these extremes are clusters that look like dumbells, with two large, round clumps. In these we could be seeing the final step in the merging together of galaxies into clumps of two, then four, eight, sixteen, or more galaxies, until there remain only two subclusters, which will eventually merge into a uniform, Coma-type cluster.

"Before *Einstein* no one knew for sure how typical Coma was and whether clusters had been around for a long time," Giacconi said. "Now we know that clusters are just now collapsing, and we know it because of x-ray observations."

After *Einstein*, Giacconi had plans to build a bigger and better x-ray observatory. This Advanced X-ray Astrophysics Facility, or AXAF, as it is called, would be a permanent observatory. It would be launched by the Space Shuttle, could be maintained in orbit, and retrieved for major overhauls. Giacconi's proposal was that AXAF would be run as a national observatory, much in the manner of the Very Large Array Radio Observatory; he further proposed that a national x-ray astronomy institute similar to the National Radio Astronomy Observatory be set up for this purpose. Because of the successes of *Einstein*, Giacconi and his colleagues—especially Harvey Tananbaum, who ably succeeded Giacconi as director of the High Energy Astrophysics Division at the Harvard–Smithsonian Center for Astrophysics—have been able to marshall widespread support in the scientific community for AXAF. But government funding for the project, whose cost would approach a billion dollars, was slow in coming. And the idea of a national x-ray astronomy institute, which would, it seemed, inevitably be directed by Giacconi, did not exactly sweep through the astronomical community like a prairie fire.

So, without a big project in the works, Giacconi had to settle down to being a Harvard professor, something that ill-suited his dynamic personality. "I'm miscast," he told us once during those years. "I'm not a professor; I'm a scientist and a manager."

Not surprisingly, when he was offered the challenge of directing the newly formed Space Telescope Science Institute, he accepted.

THREE

Lines and Spaces:
Gamma Ray Astronomy

_13

Gamma radiation is simply the continuous extension of the electromagnetic spectrum beyond x-ray energies; there is no clear boundary between the two. In terms of physical processes, gamma radiation takes us inside the atomic nucleus; it was through the radioactive decay of the nucleus that gamma radiation was discovered near the turn of the century. Nuclear gamma rays have energies that range from a few hundred thousand to a few million electron volts (an electron volt is the energy acquired by an electron moving through a potential difference of one volt, a terminology that has carried over from the way in which electrons were discovered), or a few hundred times greater than the energies of the x-ray photons detected by the *Einstein* x-ray observatory.

Gamma ray astronomy has not developed as rapidly as x-ray astronomy, in part because of difficulties inherent to the detection of gamma rays, and in part because cosmic gamma rays are rare. On the one hand, gamma rays are so energetic that they would shoot right through the relatively light gas-filled counters used in the development of x-ray astronomy. Therefore, massive detectors are required to measure gamma rays. On the other hand, like x-rays, gamma rays interact strongly with the atmosphere, so these detectors have to be lifted high above the surface of the earth in balloons or on satellites; this is an expensive proposition, and only relatively modest experiments have been flown so far. Once the detector gets into space, the problems of the gamma ray astronomer are just beginning. Our atmospheric blanket protects us not only from ultraviolet radiation, x-rays, and gamma rays, but also from a deadly bath of high-energy cosmic rays. These charged particles produce a spurious gamma ray background or static that the gamma ray astronomer must cope with. Finally, probably because of the great energy of gamma ray photons, stars and galaxies produce them sparingly. The instruments flown so far have detected about one photon every minute from the strongest sources. In contrast, the *Einstein* x-ray observatory could collect a hundred or more photons in a minute from many sources. Nor can gamma ray astronomers hope to improve their lot, as have x-ray astronomers, by building a mirror to focus gamma rays. It is difficult to do with x-rays, and impossible with gamma rays, because of their high energy.

But astronomers, like fishermen, are optimists, so a number of talented and incredibly persevering scientists accepted the challenge of gamma ray astronomy. Many of them were intrigued by the possibility of directly observing nuclear processes in a cosmic setting. Allan Jacobson was such a person.

Jacobson traveled a circuitious and colorful route to his career in astrophysics. An imposing man at 6 feet 4 inches and 200-plus pounds, he has a magnificent bass voice that so impressed the musician Gordon Jenkins that he arranged for Jacobson to cut a record for RCA. At age twenty-four, when many future astronomers are hard at work on their doctoral theses, Jacobson was pursuing a career in show business.

Bud, as his friends and colleagues call him, was born in 1932 in Chattanooga, Tennessee, and lived there for eighteen years. During that time, he had no inkling whatsoever of a future career in astronomy.

"I was a terrible student in high school. I simply wasn't interested," he recalled, as we sat around a table outside a cafeteria at the Jet Propulsion Laboratory in Pasadena on a warm day in March. "The only course that I did well in was plane geometry. I don't know why, but I took to it, and was able to do all the homework during class hours. I never cracked a book at home. I avoided school as much as possible. I always considered myself somewhat of a dunce."

After graduating from high school with a C− average, Jacobson joined the Air Force. "I blossomed in the Air Force," he said. "My aptitude scores were quite high, and within two months I found myself leading a large group of men, drilling them and so forth. Of course I was picked for that because I could do 'sound-off' like Vaughan Monroe," he laughed. "So, I was fairly successful in the Air Force and toyed with making a career of it. The reason I got out was that I had aspirations for a career in show business."

"I had teamed up with a young comedian, Billy Mears, while I was in the Air Force. We did a lot of night club shows. We did some 200 night club shows in Japan. Our plan was that, after getting out of the Air Force, we would meet in California, where he lived, and kick off our show business career. I came to California in February 1955. We had some minor successes; we were on the Spade Cooley show." Though they have remained good friends over the years, the Mears and Jacobson show business partnership was short-lived.

"We went our separate ways," he continued. "I had written some songs, and I took them to a publisher and sang them for him. He liked my voice a lot better than he liked my songs. He got me an appointment at RCA Victor to audition for Henry René, Gordon Jenkins, and one of the RCA vice presidents." A week later Gordon Jenkins called, with an offer to do an album.

"I agreed, readily," Jacobson recalled. "What I did was Rogers and Hammerstein's 'Pipe Dreams.' The problem was that the show was a flop on Broadway. One of Rogers and Hammerstein's few flops," he laughed. "I think it ran only six weeks, so RCA decided not to release the album."

During this time Jacobson had been having second thoughts about a show business career, even before "Pipe Dreams" turned out to be a pipe dream. "I was meeting a lot of very flaky people," he said, "and decided that it was not the life for me."

While trying to launch his show business career, Jacobson worked at a variety of jobs. One of these was for the Los Angeles Police Department, as a clerk in the Records and Identification section. "I really disliked working there, and began to fish around for alternatives. I was always interested in art. Drafting was another course that I had done well at in high school. So, I bought drafting tools and a book, and made drawings." He took these drawings around to prospective employers. "Not many people would talk to me because of my lack of experience, but finally one guy did, at Pacific Semiconductors, a company that is now part of TRW. He hired me on probation."

"I liked the work very much," he continued, "and I was soon promoted to mechanical designer." It was at this time, in 1957, that Jacobson began to think about a career in engineering. He enrolled in night school at Los Angeles City College. It was there that he took his first course in physics and found, somewhat to his surprise, that he liked it. He changed his major to physics. He soon realized that he needed to go to school full time. He and his wife, Edith, whom he had married in 1956, saved as much money as they could from their jobs. By 1959 he was able to quit work and attend UCLA full time, with the help of the GI Bill.

"They weren't going to let me into UCLA because I had such a poor scholastic record," Jacobson chuckled. But because he was a veteran, they let him take a battery of timed aptitude tests. After the tests, he met with a counselor. "He asked me if I had ever taken a course in speed reading. I said, 'No,' and he said, 'Well, you were not supposed to finish each of these tests, but you finished all of them.' " On the basis of the tests, he was admitted. Three years later he graduated Phi Beta Kappa.

Jacobson applied to the University of California, San Diego, graduate program in physics. "My intention was to go to UCSD and study solid-state physics. But when I got there, I found that Walter Kohn, whom I wanted to study under, had his full complement of graduate students. So, I wound up working for Norman Rostoker in theoretical plasma physics." He quickly realized that theoretical plasma physics was not the field for him. "I started looking around for someplace to land. The way I got into astronomy," he laughed, "was that Larry Peterson had an opening." Peterson had recently joined the faculty of UCSD to initiate a program there in x-ray and gamma ray astronomy and needed graduate students.

His choice of a thesis project, however, was less arbitrary. "One of the things that attracted me to physics at UCLA was spectroscopy. I had a penchant for art. The first career I ever wanted to have was as a cartoonist. I loved optical spectroscopy. I loved the colors, the sharpness, the lines. There was something that was really attractive to me about the physics behind it and the visual effects. So, when I was working for Larry, I was not happy with the way gamma ray spectra looked in scintillators."

In a typical scintillation counter, or scintillator, a crystal of sodium iodide, which looks much like rock salt, is used to absorb incoming gamma rays. The energy of the gamma rays is converted into the energy of moving electrons, which is in turn converted into photons of visible light. These photons produce a flash of light, or scintillation, which is picked up by optical sensors and converted into an electrical pulse. The height of the pulse is proportional to the energy of the gamma ray; so by comparing the pulses produced by different gamma rays, you can compare these energies. With this method it is possible to distinguish a gamma ray with an energy of 500 thousand electron volts from one with an energy of 450 thousand electron volts or 550 thousand electron volts, but it is not possible to distinguish between gamma ray energies of 500 thousand and 511 thousand electron volts, for example. This intrinsic coarseness in energy resolution will cause sharp peaks in the gamma ray spectrum to be broadened into smooth hills.

With detectors that use semiconducting crystals such as germanium, on the other hand, any strong emission lines that may exist in a gamma ray spectrum stand out like the Himalayas. A semiconductor is a material that can be either an electrical conductor or insulator, depending on the situation. An incoming gamma ray produces electron-ion pairs, temporarily allowing the semiconductor to conduct electricity if the opposite sides of the crystal are at different electric potentials. The strength of the electrical pulse produced by an incoming gamma ray is proportional to the energy of the incoming gamma ray. One difference between a scintillating crystal such as sodium iodide and a semiconducting crystal such as germanium is that in the sodium iodide one detects visible photons produced by the incoming gamma rays, whereas in the germanium, one detects electron-ion pairs. This leads to the second, crucial difference. It takes much less energy for a gamma ray to produce an electron-ion pair in germanium than to produce a visible photon in sodium iodide. For example, a 500 thousand electron volt gamma ray would produce about a thousand visible photons if it were to be absorbed by a scintillator; a gamma ray of the same energy would produce about 140,000 electron-ion pairs in a germanium crystal. This makes the gamma-ray-induced pulses from the germanium crystal much sharper than from the scintillator, which means that much sharper peaks can be resolved in the gamma ray spectrum.

It was this sharpness, reminiscent of the sharp, distinct colors of the optical spectrum, that Jacobson sought to capture. "I found myself drifting toward spectroscopy," he recalled. "When germanium detectors were developed, it was a natural for me. I wanted to get that fine structure; that's what drove me."

Why did Jacobson, or anyone else, care so much about the fine structure of cosmic gamma ray spectra? Because it could have a direct bearing on the validity of one of the central tenets of modern astrophysics, namely, that matter evolves. All the elements heavier than hydrogen are believed to have been built up from primordial hydrogen, the lightest element. Observations and theoretical calculations indicate that most of the helium, the next heaviest element, was produced in the first few minutes of the Big Bang, the explosion from which the universe as we know it is thought to have originated. The third, fourth, and fifth heaviest elements—lithium, beryllium, and boron—are thought to have been produced by collisions of cosmic rays with heavier elements in the gas between the stars.

Where did the heavier elements originate? In stars. Nuclear reactions inside stars such as the sun, in even heavier stars, and in stellar explosions (supernovae) thought to be capable of producing all the elements from carbon to uranium. So the carbon in our muscles, the iron in our blood, and the calcium in our bones were produced in stars, along with all the other elements heavier than carbon in our bodies, on the earth, and in the rest of the universe. In the words of Harlow Shapley, "We are brothers to the boulders and cousins to the stars."

The basic idea that the elements are synthesized in stars was first put forth by Fred Hoyle in 1946. In the course of the next decade, Hoyle began a collaboration with William Fowler of the California Institute of Technology which culminated in a classic paper by Fowler, Hoyle, and Margaret and Geoffrey Burbidge of the University of California, San Diego. Published in the *Reviews of Modern Physics* in 1957, this paper concluded, with the characterstic bravado that has made Hoyle such an influential and controversial figure in modern astrophysics, that "we have found it possible to explain, in a general way, the abundances of practically all the isotopes of the elements from hydrogen through uranium by synthesis in stars and supernovae."

Alistair Cameron of Harvard University presented essentially the same conclusions in a paper published the same year, but Cameron was not Hoyle, and he published in a more obscure journal; consequently, his work attracted little notice at the time. Given Cameron's work, it is clear that the theory of the synthesis of the elements in stars would have eventually gained the widespread acceptance it enjoys today without Fowler and Hoyle. But it might have taken much longer.

If Hoyle was the messiah of stellar nucleosynthesis, then Fowler was its Saint Paul. Though "Saint" is hardly a title that fits the gregarious Fowler, he did use his prodigious energy and enthusiasm to spread the gospel of stellar nucleosynthesis. In the years that followed, he and his colleagues laid the experimental foundations for the theory through the measurement in the laboratory of reaction

rates for crucial processes, and they made numerous theoretical calculations that illustrated the plausibility of the theory. For these
efforts, Fowler was awarded a Nobel Prize in physics in 1983.

It is an elegant, well-developed theory supported by an impressive
body of circumstantial evidence, but little direct evidence. The remnants of supernovae were widely believed to be the best place to
look for the "smoking gun" of the stellar synthesis of elements, for
two reasons. First, the insides of the star are ejected into interstellar
space in a supernova explosion; therefore, the environs should be
enriched in the elements synthesized inside the star. Second, the
explosion itself is believed to produce many different types of elements, some of which are radioactive. The decay of radioactive elements produces gamma rays; thus, the detection of gamma rays from
a supernova would be convincing proof of the theory of stellar nucleosynthesis.

It was Jacobson's goal to provide such proof. As part of his doctoral
thesis project, he built a germanium gamma ray detector. The system
consisted of a germanium detector having a diameter roughly equal
to that of a half dollar and about twice as thick, cooled by liquid
nitrogen—germanium detectors will not work unless they are cooled
below about −200° Celsius—and packaged with the necessary electronics to retrieve the data and guide the balloon which would carry
the detector about 20 miles above the surface of the earth. With this
system Jacobson made the first observation of an object outside the
solar system with a germanium detector. He pointed the detector
toward the Crab Nebula, the remnant of a supernova explosion, in
hopes of detecting gamma ray lines from radioactive elements thought
to have been produced in abundance in the supernova. Evidently
the theorists had been overly optimistic in their predictions of the
amount of radioactivity produced by a supernova. Jacobson was
disappointed to find that the gamma ray spectrum was a smooth
continuation of the x-ray spectrum, with no hint of lines. The experiment had a measure of success, though, because he had demonstrated the feasibility of using germanium detectors.

After obtaining his Ph.D. in 1968, Jacobson stayed on at the University of California at San Diego for a year, but soon he wanted to
branch out on his own. He left to take a position at NASA's Jet
Propulsion Laboratory in Pasadena.

_14

In the early 1970s gamma ray astronomy began to demonstrate its usefulness for investigating a wide range of phenomena. James Kurfess of the Naval Research Laboratory had used gamma ray detectors carried aloft by high-altitude balloons to show that a large fraction of the energy radiated by the rotating neutron star in the Crab Nebula is given off in gamma rays with energies approaching a million electron volts. Ian Strong, Ray Klebesadel, and Roy Olson of Los Alamos Scientific Laboratory used data gathered from a network of United States Defense Department satellites to establish the existence of a new astronomical phenomenon: gamma ray bursts that occur at random from random parts of the galaxy. Understanding the nature of these bursts, which are believed to originate on neutron stars, remains a major unsolved problem in astronomy.

Balloon experiments by Neil Johnson of the Naval Research Laboratory and Robert Haymes of Rice University provided evidence for the production of antimatter in the central regions of the galaxy. In collisions between energetic particles, some of the energy of the collision can go into the production of new particles and antiparticles. An antiparticle is a subatomic particle that has the same mass as the particle to which it corresponds but an opposite charge, among other properties. For example, an antiproton has the same mass as a proton but a negative charge. An antielectron—or positron, as it is called—has the same mass as an electron but a positive charge. When a particle encounters its antiparticle, the particle and antiparticle annihilate each other and their mass is converted into another form of energy. For example, when an electron and a positron meet, they produce gamma rays. The details of the gamma radiation depend on a number of things, such as how fast the electron and positron were moving when they collided, but in general it is sharply peaked around an energy of 511 thousand electron volts. Johnson and Haymes found a broad peak in their scintillation counter data at this energy, and subsequently Marvin Leventhal of Bell Telephone Laboratories and his colleagues used a germanium detector to confirm the existence of a sharp peak at 511 thousand electron volts.

Earlier, William Kraushaar and his colleagues at the Massachusetts Institute of Technology had used a small gamma ray detector aboard one of NASA's Orbiting Solar Observatory satellites to show that the disk of the galaxy, especially the region toward the center of the galaxy, is a strong source of gamma rays having energies greater than a few tens of millions of electron volts. Above the disk, the intensity of the gamma radiation dropped off, but not to zero: a diffuse background glow of gamma rays remained. This background radiation is, like the x-ray background and the microwave background radiation, uniform over the sky. Its origin is still not understood. A popular idea is that, like the x-ray background, it is not truly diffuse but is produced by a large number of distant, unresolved sources such as active galactic nuclei and quasars.

In late 1972 NASA's second Small Astronomy Satellite, *SAS-2* (*Uhuru* was *SAS-1*), carried into orbit a small gamma ray telescope, with an effective area roughly equal to that of a three-inch optical telescope. This mission, under the direction of Carl Fichtel of NASA/ Goddard, made a gamma ray map of the sky during its seven-month lifetime. This map confirmed the results of the MIT group, both with regard to the diffuse background and the concentration of radiation along the disk of the Milky Way. With *SAS-2*, the Goddard group was able to show that part of the gamma radiation from the disk of the galaxy is coming from individual sources. The Crab Nebula was confirmed as a gamma ray source, a pulsar in the constellation of Vela was found to be a bright source of gamma rays, and the x-ray source Cygnus X-3 was also detected. These results suggested that gamma ray sources are, like binary x-ray sources and radio pulsars, associated with neutron stars. The resolution of the *SAS-2* gamma ray detector was still crude by the standards of optical astronomy, however. It had an angular resolution of a few degrees; in other words, they could not resolve any details smaller than a few times the angular size of the full moon. This made it difficult to tell whether or not all the gamma radiation from the galactic plane could be accounted for by individual sources, but the observed smooth distribution of gamma rays across the galaxy suggests that a large fraction of the galactic gamma radiation is produced by collisions of high-energy protons with clouds of interstellar gas distributed along the spiral arms of the galaxy. This result demonstrated the contention of several early theoretical papers that gamma ray observations had the unique ability to study processes that generate high-energy protons in the distant reaches of our galaxy.

The results from *SAS-2* were coming in while NASA was in the

midst of restructuring its planned program of High Energy Astronomy
Observatories, the first of which was to be launched in 1976. Despite
the successes of *SAS-2*, it was not possible to include, as originally
planned, a bigger and better version of the *SAS-2* instrument. It would
be too heavy for the rockets that would be used to launch the new
missions. One gamma ray experiment was selected, however. It was
a gamma ray spectrometer, which would be constructed under Allan
Jacobson's direction at the Jet Propulsion Laboratory.

The spectrometer was essentially a bigger and better version of
the gamma ray spectrometer he had built for his thesis project. With
its cluster of four germanium detectors cooled to $-193°$ Celsius by
solid methane and ammonia, it was about a hundred times as sensitive as Jacobson's first spectrometer. "It was a long and difficult
project," he recalled. "Invariably we were over budget and behind
schedule. I was subject to a lot of pressure from the outside, and I
didn't always have total control inside, because JPL doesn't vest their
scientists with total control of the experiments. Usually the control
is put in the hands of engineering management. It was like being
caught between a rock and a hard place. One is held responsible for
holding to budget and schedule from the outside, but exercising any
power on the inside to accomplish that was very difficult."

All the HEAO experiments had difficulty staying within budget;
the main culprit, however, was not the scientists and engineers but
inflation. During the decade when the HEAO projects were constructed, the consumer price index for commodities and services
increased by more than 50 percent. "I have gone back and looked
at the numbers," Jacobson said, "and we tracked inflation. So what
was happening was that we were getting beaten, trying to stop what
was an economic pressure in the country. There was no way we
were going to overcome that pressure."

But NASA headquarters kept applying pressure of their own, to
keep the project on budget and on schedule. Sometimes the pressure
was heavy-handed. "They would call up and threaten to put another
experiment in my place," he said. "I don't think they were ever
serious about doing that, but there were a couple of times when they
went to some lengths to make it look like they were serious. I remember specifically they said, 'We're getting ready a cosmic ray
experiment at Goddard to take your place.' A guy who did me a big
favor at that time was Frank McDonald. On a visit to JPL he came

by to talk with me, and told me that they came out to Goddard and asked if he could make it appear that he was getting a cosmic ray experiment ready. McDonald told them not to play those stupid games, and basically cut that off."

Often during those times, Jacobson seriously considered resigning from the project. He was under tremendous pressure at work and at home, where his marriage of over twenty years was breaking up. "There was so much pressure on me during the HEAO program," he said, "that I developed a whole bank of medical problems." These problems were diagnosed as being related to the extreme tension under which he was working; so in 1978, after the instrument had been built and calibrated, he decided to take a sabbatical leave at the University of California, Los Angeles. He would still be close at hand and available for meetings, but he would be out of the pressure cooker for nine months while the instrument was prepared for launch. It helped.

"It took me several months to get over all of the symptoms and they have never come back," he said, "but soon after I went away for a sabbatical, I had to go to a meeting at TRW. The instant I walked into that meeting, all of these symptoms popped up at one time." He laughed. "I couldn't believe it."

The major source of tension on all projects of this type is how best to reconcile the internal pressures to do the best job possible and the external pressures of time and money. "It's a lot of worry," he continued, "if you're very conscientious. I wanted the thing to be right and I wanted it to be good. It was tough technology and I wanted it to work well. And everybody is trying to get you to lower the scope. Everybody's trying to get you to make it easier for them. I fought that every inch of the way."

The worst crisis of the project came just four months before launch. The gamma ray spectrometer had passed vibration and thermal vacuum tests and appeared to be ready for integration onto the spacecraft with the other experiments. Then it began to experience "thermal excursions" in which the temperature in the instrument would rise some 40° higher than the design temperature of − 193° Celsius. The Jet Propulsion Lab called in a "tiger team" of problem solvers, headed by Frank Shutz and Jim Stevens. Their grim assessment after two weeks of sixteen-hour days: the instrument would have to be taken apart.

Everyone familiar with the instrument was called in, from JPL, NASA/Marshall, TRW, and Ball Brothers Aerospace in Boulder, Colorado. The sixteen-hour workdays continued through the summer. Working with all the care of a bomb-disarming team, they disassembled the instrument until they found the problem. A washer had not been squeezed flat by its companion bolt. The less-than-perfect seal had caused the thermal excursions. The seal was repaired and the instrument reassembled. By then they were getting into September, and other problems loomed. If the spacecraft was not launched before October 1, the fiscal year would end. There was no money in the 1980 budget for Atlas Centaur launches. The red tape involved with rescheduling the launch to another fiscal year could cost millions of dollars. The High Energy Astronomy Observatory program managers had the satellite at Cape Canaveral ready for launch. Two instruments designed to measure the properties of cosmic rays were on the spacecraft and ready to go. The launch vehicle was ready. Everything was ready but the gamma ray spectrometer.

As soon as the gamma ray spectrometer was reassembled, it was rushed to Cape Canaveral by the Military Air Transport Service. At the Cape, another problem appeared on the horizon. "There was a hurricane headed for the Cape at the time," Jacobson explained. "We just had time to land the plane, unload the experiment, integrate it onto the spacecraft and evacuate." The spacecraft and launch vehicle were secured against the hurricane, while the nervous scientists and engineers waited. Fortunately, the hurricane missed the Cape, and on September 20, 1979, the third High Energy Astrophysical Observatory, *HEAO-3*, was launched.

It was an emotional moment. "It was a very tough program. It was tough techically to accomplish. I feel very proud that we did it. It was a highly successful experiment with a lot of difficult technologies. I felt so strongly about it that when I watched it launch I started to cry. When the thing took off I found myself crying like a baby." He laughed at the memory.

After the launch, Jacobson went to Goddard, to prepare for the arrival of the first data. "I had in mind at the time that I was going to see if there was any net half MeV [500 thousand electron volts] flux coming from the galactic center. But at that time we did not have a full appreciation of all the systematic effects in a gamma ray experiment."

Their problem was exacting and wearisome. The gamma rays they sought—those coming from the galactic center, or from interstellar space, or from stars—came not in a flood but in a trickle—one every few minutes. During the same few minutes, cosmic rays or charged particles trapped in belts around the earth peppered the instrument, producing a thousand or more unwanted gamma rays. Before Jacobson and his team of James Ling, William Mahoney, Guenther Riegler, and William Wheaton could find the gamma rays they wanted to study, they had to learn as much as possible about the unwanted gamma rays. What causes them? How many are there? What is their energy? Do their numbers change with the direction the spacecraft is pointed or with its position in orbit? Their situation was analogous to that of a diamond miner, who must muck through tons of ore and be able to identify all the different low-quality stones before he can find the precious gems.

For two frustrating years, while many people wondered, some of them out loud, where the results were, Jacobson's group carefully and patiently sifted through the data and sorted out numerous systematic effects. "Ever since the early days of gamma ray spectroscopy, there's been a credibility problem," he explained. "I have been very sensitive to that, and have been super-careful to the point where I have often been accused of being much too slow to publish. I spend a lot of time to be as certain as reasonably possible that we're not making a mistake. If there's anything that characterizes my research, I would like it to be high technical competence and high credibility. Certainly I can't say enough about my group. They're all topnotch people, all extremely competent and careful."

One of the many effects they had to contend with was the radioactivity induced by collisions of cosmic rays with the instrument. "The material becomes radioactive," he explained. "We had some 120 to 140 lines in the background from internal activity of the detector materials. The thing that makes a high-resolution spectrometer work well is not that it is subject to more or less background than a scintillator, but that you can get a handle on it. You can see each one of those lines. It stands out so clearly that it is identifiable and hence the process that generated it is identifiable and hence you can deal with it. In a scintillation detector, with moderate to poor resolution, all you know is that the background is changing continuously, and you have a very difficult time getting a handle on the systematics. Everything changes with time. The magnetic field of the earth causes a modulation in orbit of the cosmic rays which modulates all the background, and then you go through the South Atlantic Anomaly [a region where the concentration of trapped charged particles is especially high] . . . It's a very difficult problem and I think we've done a good job in unraveling it. We brought out some very good results."

Among the first results they brought out were observations of the electron-positron annihilation line at 511 thousand electron volts, the feature Jacobson had hoped to pull out of the data two years earlier. They found that the line was variable over a period of about six months. This means that the source of the radiation can be no larger than half a light year, a finding which supports the idea that the radiation is coming from the vicinity of a massive black hole in the center of our galaxy.

"I am most pleased with the aluminum 26 discovery," he said. The story of aluminum 26, a radioactive isotope of aluminum, illustrates how our picture of the universe is an interlocking mosaic of ideas and observations, how each small piece is important, how changing one small piece can have an effect on the whole picture, and how scientists, like everyone else, sometimes do the right thing for the wrong reason.

The 26 in aluminum 26 refers to the combined number of protons (13) and neutrons (13) in the aluminum nucleus. Certain combinations of protons and neutrons are stable and others are radioactive, that is, the nucleus spontaneously changes into the nucleus of another type of atom with the radiation of energy. For example, aluminum 27 is stable, whereas its sister isotope, aluminum 26, is not. It spontaneously decays to magnesium 26, a stable nucleus with 12 protons and 14 neutrons. The most probable by-product of this radioactive decay is a gamma ray with an energy of 1.809 million electron volts and a positron.

It is impossible to say exactly when the radioactive decay of any given unstable nucleus will occur. It is analogous to predicting when a given kernel of popcorn will pop. You cannot say exactly when it will happen, but if you have a large number of kernels, you might be able to say that after a certain length of time, half of the kernels will have popped. This would be the half-life of the kernels, and it might vary from one brand or type of popcorn to another, or even with the type of popper. Unlike the kernels of popcorn, though, the half-life of a radioactive nucleus does not depend on external conditions. It follows an internal clock peculiar to each type of nucleus.

The half-life of aluminum 26 is 740,000 years. This means that if we were to create a million aluminum 26 nuclei in a particle accelerator today, half of them would decay in the next 740,000 years, and half of them would be left. After another 740,000 years, half of the remaining aluminum 26 nuclei would decay, leaving only one quarter of the original sample, and so on. After ten million years only a small fraction of the original sample would be left. The rest of it would have changed into magnesium 26.

In the early 1970s Gerald Wasserburg and his colleagues Typhoon Lee and D. A. Papanastassiou at Cal Tech found a surprisingly large amount of magnesium 26 in aluminum-rich inclusions in meteoritic material. Since magnesium 26 is produced by the radioactive decay of aluminum 26, their finding implied that a large amount of aluminum 26 must have been present in the material from which the meteorite formed. Other lines of evidence indicate that meteorites formed at about the same time as the sun and planets (about 4.5 billion years ago) and that they originated in the same cloud of gas and dust. Therefore, Wasserburg and colleagues concluded that the cloud which formed the solar system must have been enriched in aluminum 26. Since the half-life for decay of aluminum 26 is slightly less than a million years, the enrichment apparently happened less than a million years or so before the solar system collapsed into its present shape. The most likely source of this radioactive aluminum seemed to be a supernova explosion.

The chance that such an explosion would occur just about the time when the solar system was formed is very small, unless the two events were connected. Therefore, suggested Cameron and James Truran of the University of Illinois, the supernova might have actually triggered the formation of the solar system. This attractive idea soon gained widespread popularity, evoking the image of the creation of a new star from the destruction of an old one.

David Arnett of the University of Chicago suggested an observational test of this hypothesis. If aluminum 26 were produced in large enough quantities in a supernova explosion to explain the peculiarities found in the meteorites, then perhaps the gamma radiation from the decay of aluminum 26 could be detected. The time between supernova explosions in our galaxy is thought to be about 30 years, so the debris of about 25,000 explosions could accumulate within the 740,000 year half-life of aluminum 26. About the same time, two scientists who were exploring the prospects for gamma ray astronomy, and who were apparently unaware of the Cal Tech results on meteorites, reached essentially the same conclusion as Arnett. Reuven Ramaty of NASA/Goddard and Richard Lingenfelter of UCLA made a theoretical survey of gamma ray lines that might be detected with the gamma ray spectrometer aboard the High Energy Astronomical Observatory or on future experiments. They concluded that the decay of the radioactive isotope aluminum 26 "is a very good candidate for producing a detectable gamma ray line."

These authors realized that broader issues than the origin of the anomalies in some meteoritic material, or even the origin of the solar system, were at stake. The detection of aluminum 26 gamma rays would, first of all, provide one of the few bits of solid evidence that the elements are produced by stellar explosions. Second, the aluminum 26 gamma rays must be coming from a radioactive sample that was produced less than a few million years ago—only yesterday in terms of the galactic time scale. Otherwise, almost all of the aluminum 26 would have decayed to magnesium 26. The detection of gamma rays from the decay of aluminum 26 would therefore provide proof that the explosive synthesis of the elements is occurring in the galaxy in our own time.

In the November 15, 1984, issue of the *Astrophysical Journal,* Mahoney, Ling, Wheaton, and Jacobson published a paper describing the detection of a gamma ray emission from the disk of the galaxy in a narrow line at 1.80 million electron volts, exactly the energy expected from the decay of aluminum 26. The line was weak—they detected only three of the desired photons for every 30 minutes of observing, on the average—but because of their careful attention to the details of the background gamma radiation, they were confident of its existence.

The intensity of the line is several times stronger than even the most optimistic estimates for supernovae, and most researchers now seem to agree that another source of aluminum 26 must be sought. The most popular sources at present are novae, which are believed to be thermonuclear outbursts that occur on the surface of white dwarf stars. A nova, unlike a supernova, does not disrupt the entire star; it blows off only a fraction of a percent of the mass of a white dwarf. But because novae occur a thousand times more frequently than supernovae, their accumulated contribution of aluminum 26 can outweigh that of supernovae.

Ironically, the reason for believing that gamma radiation from aluminum 26 could be observed no longer seems valid, but the radiation has been observed anyway. Nor is it necessary to assume that a nearby supernova explosion triggered the formation of the solar system and enriched the primordial solar cloud in aluminum 26. The interstellar medium from which the solar system collapsed is continually being enriched by scores of novae every year, and some other event (the passage through a spiral arm of the galaxy, for example) could have triggered the collapse. It is a commonly repeated theme in astronomy: a theoretical idea is taken seriously and leads to an observation that proves the idea wrong; but the observation, in turn, leads to new insight and understanding.

_15_

The next major step forward in gamma ray astronomy will be NASA's Gamma Ray Observatory. It is scheduled for launch by the Space Shuttle in 1987. The 15-ton observatory will carry four instruments designed to explore gamma ray sources over a very wide range of energies, from about a 100 thousand electron volts to 30 billion electron volts. The instruments were chosen to complement one another and to provide coordinated observations of the entire gamma ray spectrum with a better sensitivity than any other previous mission.

James Kurfess of the Naval Research Laboratory is the principal investigator for the Oriented Scintillation Spectrometer Experiment (OSSE), an array of four scintillation detectors which will cover the energy range from 100,000 to 10 million electron volts. The Imaging Compton Telescope is a device which covers the range from 1 million to 30 million electron volts and uses two sets of scintillation detectors as a means of determining the direction of the incoming gamma rays and of mitigating the gamma-ray background problem. The principal investigator for this experiment is V. Schonfelder of the Max Planck Institute in Munich. The Energetic Gamma Ray Telescope, a larger and much improved version of the *SAS-2* instrument, will cover the high-energy range from 20 million electron volts to 30 billion electron volts; the principal investigators are Carl Fichtel of NASA/Goddard, Robert Hofstader of Stanford University, and Klaus Pinkau of the Max Planck Institute. The Burst and Transient Source Experiment will continuously observe the full sky for gamma ray bursts or other transient sources of gamma radiation; Gerald Fishman of NASA/Marshall is the principal investigator.

Conspicuous by its absence is a high-resolution gamma ray spectrometer of the type that flew on the High Energy Astronomy Observatory. The Gamma Ray Observatory, as originally planned, had such an instrument, the Gamma Ray Spectrometer Experiment (GRSE), which was to involve a broad collaboration between scientists at five institutions, including Allan Jacobson's group at the Jet Propulsion Lab. Because of escalating costs, the spectrometer was eliminated.

"NASA made a horrible mistake in throwing the high-resolution spectrometer off of GRO," Jacobson said. "The story is that they can do everything with the scintillation spectrometer. I don't believe it, because, remember that Haymes [and Johnson] observed the galactic center half MeV [million electron volts] feature, but never could get the energy quite right. No one considered it credible; it wasn't until Marvin Leventhal observed it with a high-resolution spectrometer . . . As far as extra-solar-system gamma rays are concerned, the existence of lines has never been established with scintillators."

The problem, as always in gamma ray astronomy, is the unwanted background radiation. "I'll give you an example of the problems with scintillation detectors," Jacobson continued. "*HEAO-1*-4 [a high-energy x-ray and low-energy gamma ray detector aboard the first HEAO spacecraft]. It covered the same energy range as *HEAO-3*-3 [Jacobson's experiment]. They have never reported the half MeV line from the galactic center. They know about the aluminum 26, but haven't reported it in their data. Where are all these results? . . . That's what concerns me about OSSE. I hope it does the job. I know there are some jobs it can't possibly do. One should be able to see the mass motions of the aluminum 26 in the Doppler shifts." During its lifetime of roughly a million years, the aluminum 26 will become mixed with interstellar clouds. By studying the subtle shifts in the energy of the gamma ray line, it would be possible to study the motions of these clouds. "You can't do that with scintillators, and that has got to be key information that relates to the source."

Jacobson was not optimistic about his future in gamma ray astronomy. "At this moment I feel like somebody who has done a very successful experiment and has no place to go," he said. "I don't see any more opportunities in the future. Surely somebody has to measure the distribution of aluminum 26, find the source, map it. I have my reservations about whether GRO can do an adequate job of this. There are no other opportunities." Nor was he encouraged by another development with regard to the GRO.

Gerry Fishman, the principal investigator of the Burst and Transient Source experiment on GRO, explained this to us, when we asked how long the mission is scheduled to last. "Two years, minimum," he said. "They have approval for refueling capability, so that the shuttle can go up and refuel the hydrazine tanks. That could extend the mission indefinitely." That is good news for Fishman and the other scientists who have experiments on GRO, but not for those on the sidelines.

"That will make the opportunities for developing new instrumentation even scarcer," Jacobson said. "They are moving to bigger and bigger instruments and there are fewer and fewer of them. That's a necessity," he conceded. "You have to do it in a big way now. I just don't see any opportunities, and I'm getting very discouraged over that."

The thought of quitting the field has occurred to him more than once recently. "I find that I am doing more exciting work in computers than I am in gamma ray astronomy." Jacobson consults on a regular basis for some of the major computer software companies. "We have a lot of good results, but that's going to die out, and the funding is going to gradually die out, and then I will be left with a group of highly trained individuals who probably know more about low energy gamma ray astronomy and the systematics involved than anyone else in the world. It's a large group. What are they going to do? Where are they going?"

He reflected on this a moment, rattling the ice cubes in the paper cup in front of him. "It's a science that's interesting. I would have thought that having such a successful flight and having such good results would give me leverage, but it hasn't."

We asked if he had any insight into why this was, why the spectrometer and not some other experiment had been bumped off the Gamma Ray Observatory. "I don't know," he said. "Larry [Peterson] was the PI [principal investigator] and I wasn't always deeply involved in what happened, so I'm not sure how it happened. I just don't know. All I know is that the story is that it costs too much money, but in fact it is a small amount of money compared to the whole mission. So, I don't know what happened. A lot of people got together and said, 'Well, you can do it all with a scintillation counter, so what do you need with a solid-state detector?' This, incidentally, was after they had used the solid-state detector as the major selling point for GRO! That had been the star experiment when they were selling the GRO," he laughed and shook his head.

Jacobson is upset by the GRO setback, but it would be a mistake to assume that it has made him an unhappy man. On the contrary, he says that he is enjoying himself immensely, in a large measure because of his work designing computer software. "I love graphic design," he said. "My first love was art, or design, and now, finally I can do it and get paid for it. I'm having a great time." He also finds time to keep up with another of his early loves, music, and if pressed will bring his banjo to parties and sing some of the old favorites, such as "Muhlenberg County" and "Logger Lover," that made him famous among the graduate students and their families twenty years ago.

Nor has Jacobson given up on gamma ray astronomy. In spite of the setback on GRO, the limited prospects for future experiments, and his comments to the contrary, he continues to look to the future, and to work on better gamma ray detectors. "I must say," he smiled, "I hang on tenaciously. I went to NASA with a proposal for liquid argon and liquid xenon counters for gamma rays and they have given me a program to do that. It's not as big as we need, but in fact we have our first liquid argon gamma ray detector working in the laboratory now." His goal is to get around a major obstacle that stands in the way of constructing better gamma ray spectrometers—the cost of making ultrapure samples of germanium to use in the detectors.

"There has always been a limitation on what you can do with germanium detectors," he explained. "Each detector is only so big, so the way you make big instruments it to put in more and more and more of them, and that complexity can kill you after a while. You have reliability problems, cost problems, and so on. Ultimately, you want to get a huge detector, but you can't do it with germanium. As far as I was concerned, the GRO was the end of the line anyway for germanium detectors. I've always been looking for something else and right now this seems to be the best bet. With a liquid xenon detector you get a material that is as absorptive and has the same efficiency as germanium, but you can make it in large volumes. In principle, there's nothing wrong with making a boxcar-sized detector. There are all sorts of applications for such a detector."

While Jacobson watches from the wings and waits for the time, probably a decade away, when he may take the stage again with some newly developed detector, the GRO scientists are immersed in the demands of the present, as they race toward a launch in the near future. One of them knows how Jacobson feels to be left out of the action. Carl Fichtel, one of the pioneers of gamma ray astronomy, was bumped off the HEAO project in 1973; by the time the GRO flies it will have been fifteen years since his last major project.

When Fichtel first arrived at NASA's Goddard Space Flight Center in the fall of 1959, he was fresh out of graduate school at Washington University in St. Louis. He had done a thesis on cosmic rays and had come to Goddard primarily to use the sounding rockets that were available there to continue his research. But like a number of his colleagues, he was intrigued by the possibilities of x-ray and gamma ray astronomy.

"There was a lot of speculation at meetings," he remembered. "Based somewhat on the optimistic predictions of theorists, we started out with some balloon experiments." They were not successful. "It became obvious that one was going to have to go to bigger instruments and put them in space . . . I think that was where we really broke with everybody else. They were still trying to do something with balloons."

Fichtel and his group at Goddard were interested in high-energy gamma rays, ones with energies greater than a few tens of millions of electron volts. They began in late 1963 to develop a detector for these gamma rays that could fly on a satellite. In 1966 they had a successful balloon flight of their instrument. In the same year they submitted a proposal to NASA to fly one of their detectors on a small satellite. The result was *SAS-2* which had a successful but short flight from November 1972 until June 1973, when a capacitor failed. "It seems to have been one of those unfortunate random failures," Fichtel said philosophically.

Not long after the premature termination of *SAS-2*, Fichtel received more bad news. His proposal to fly a larger and more sophisticated high-energy gamma ray detector aboard the High Energy Astronomy Observatory, which had been formally accepted in 1970, had been left off the revised missions. The next mission after those was a decade or more away. Undaunted, he set to work, in his laboratory and on the committees, the astronomical equivalent of the "old boy" network.

"The high-energy astrophysics community had a working group," he recalled, "and there was generally the feeling, particularly with the restructuring of HEAO, that one ought to develop new programs to proceed." The general consensus of this working group was that in the 1980s there should be separate x-ray, gamma ray, and cosmic ray observatories. "The x-ray people felt that the Advanced X-ray Astrophysics Facility was the next step for them, and it was not really ready to proceed. The cosmic ray community had some difficulty getting organized, but it was generally agreed that the gamma ray community was in a position to move forward with an observatory." So, a gamma ray observatory was put first in line, for a projected 1984 launch. The working group's report was blessed by various other committees, and in 1981 it was given the official "new start" status by NASA. In the process the launch got pushed back to 1988.

We asked Kurfess, Fishman, and Fichtel, three of the principal investigators, to comment on some of the problems they would like to see the Gamma Ray Observatory solve.

"One of the big issues is to pin down sites from the origin of the elements," Kurfess said. "There has been some progress on that outside the gamma ray region, but I think direct confirmation of production of heavy nuclei during supernovae and novae will be important." We asked about the sensitivity of the scintillation counters to gamma ray lines, in view of Jacobson's misgivings. He explained that the Oriented Scintillation Spectrometer Experiment would be able to detect sources of gamma ray lines that are 50 times fainter than the source at the center of the galaxy; that line, as well as the aluminum 26 line from the disk of the galaxy, will stand out clearly. The experiment is designed so that four different detectors can look at different parts of the sky. This arrangement makes it possible to monitor the background radiation continually, so that it can be reliably subtracted from the measurements of the cosmic gamma ray source under observation. Because of this design, Kurfess is optimistic about the prospects for studying gamma ray lines. "We should be able to study a lot of lines in detail, and to see lines from young supernovae as far away as the Virgo Cluster [about 50 million light years distant]," he said. Prominent lines will come from radioactive iron, cobalt, nickel, possibly titanium, and of course aluminum 26. "An important thing will be to map out the aluminum 26 line around the galaxy and determine if it comes from novae, supernovae, or red giants."

Another important problem to be tackled by the Gamma Ray Observatory is that of the nature of the gamma ray bursters. These enigmatic bursts can occur any time at any place in the sky. Since they typically last only a few seconds, the problem of understanding them is analogous to that of understanding the nature of fireflies from observing random flashes of light in a meadow at twilight. In spite of this, remarkable progress has been made, and most astronomers who have studied the problem believe that the bursts are produced by neutron stars. But they aren't sure, and they are even less sure how the neutron star manages to produce the bursts.

Fishman is hopeful that his Burst and Transient Source Experiment on the Gamma Ray Observatory will provide the data necessary to solve the problem. "We should be able to observe hundreds of bursts," he said. "We will be able to find out whether the same object produces repeated bursts, and for the strong sources, measure the time variations in the intensity and spectrum of the gamma radiation. This should allow us to determine whether or not there are different classes of bursters, and what the mechanism is for producing the gamma rays, whether they're produced by thermonuclear explosions, or some other process."

In 1975, the European Space Agency launched the *COS-B* satellite, which carried aloft a gamma ray detector similar to the one aboard *SAS-2*. *COS-B* operated for over six years. During this time it confirmed and extended the results of *SAS-2*, making the most detailed map to date of the appearance of the sky in gamma rays. One of the most interesting discoveries of *COS-B* was the detection of gamma rays from a quasar. The intensity of the gamma radiation is such that 50 percent or more of the total radiation from the quasar comes out in gamma rays. Clearly, if we are ever to understand quasars and the role that massive black holes play as their powerhouse, we must know more about the gamma radiation they produce.

The Gamma Ray Observatory should make a significant step forward in this understanding. Fichtel's Energetic Gamma Ray Telescope should detect more quasars and provide an accurate estimate of the fraction of energy produced in gamma rays, an important number for the theories of quasars. The ability of the Gamma Ray Observatory to measure the gamma radiation from quasars and explosive galaxies over a wide range of gamma ray energies will be especially useful in deciding between the various theories of these objects. For example, it may be possible to determine the temperature of the gas near a central black hole, and whether or not the exotic Penrose process, in which gamma rays actually extract energy from a rotating black hole, is an important factor in quasars.

Finally, there is the question of the cosmic gamma ray background radiation. One explanation is that it is produced by many quasars or explosively active galaxies that are too faint to be resolved individually. This is essentially the same explanation as that given for the x-ray background and is the most favored one. Another explanation is based on an unorthodox cosmology which assumes that the universe is composed of equal parts of matter and antimatter, and that matter and antimatter exist separately on the scale of superclusters and galaxies—clusters of clusters of galaxies. The boundary regions between a supercluster made of matter and one made of antimatter would be aglow with gamma radiation produced by matter–antimatter annihilation. These interactions would have been more intense in the distant past when the universe was smaller and the boundary regions of matter and antimatter were compressed. The radiation from these remote epochs would have been degraded to lower energies because of the expansion of the universe.

The details of such a model have been calculated by Floyd Stecker and his colleagues at NASA/Goddard; they find good agreement with the observations. Most astronomers have not been impressed, however. The critics argue that Stecker and colleagues have assumed that the gamma ray background is produced by matter–antimatter annihilation and looked for models to fit their data. "At best," Gary Steigman of the University of Delaware has written, "such an approach can determine only if the observations are consistent with an annihilation origin. But it cannot be established that the observations demand such an explanation until all other alternatives are eliminated." Furthermore, Steigman continues, "A symmetric [showing equal amounts of matter and antimatter] cosmology that successfully meets the observational tests has yet to be proposed." The major problem of such cosmologies is that a way must be found to separate matter and antimatter before it interacts, leaving, in Steigman's words, "too little surviving matter on which to base an interesting universe."

Despite the critics, Fichtel has adopted a wait-and-see attitude. "Of all the possible explanations," he told us, "there are only two that are really viable. One is the summation of active galaxies and the other is that it is matter–antimatter interacting at the boundaries of superclusters of galaxies. I think that the GRO observations ought to be able to separate these two, and it may be one of the very few chances we have to see whether or not there is really matter–antimatter symmetry in the universe."

Whatever the outcome, Fichtel is anxiously looking forward to the flight of the Gamma Ray Observatory. When it finally does go, it will be due in no small measure to his persistence over fifteen years. Such determination and endurance have come to be viewed as part of the game. "In space astrophysics these days," he noted wryly, "it's a long time between drinks."

David Heeschen outside the National Radio Astronomy Observatory headquarters in Charlottesville, Virginia.

Riccardo Giacconi

Allan Jacobson

Carl Fichtel

Frank Low

Gerry Neugebauer

Lyman Spitzer, Jr.

John Bahcall

Charles Robert O'Dell. (*Photo by David Wolf.*)

The Very Large Array of radio telescopes. (*Courtesy of the National Radio Astronomy Observatory, operated by Associated Universities, Inc., under contract with the National Science Foundation.*)

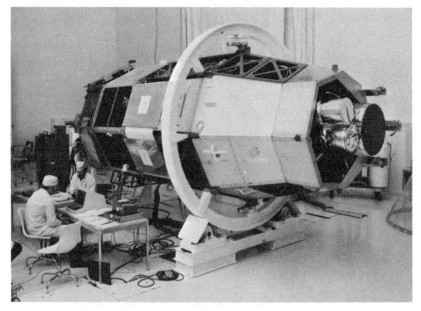

Engineers check out the *Einstein* x-ray observatory. (*Courtesy of H. Tananbaum, Center for Astrophysics.*)

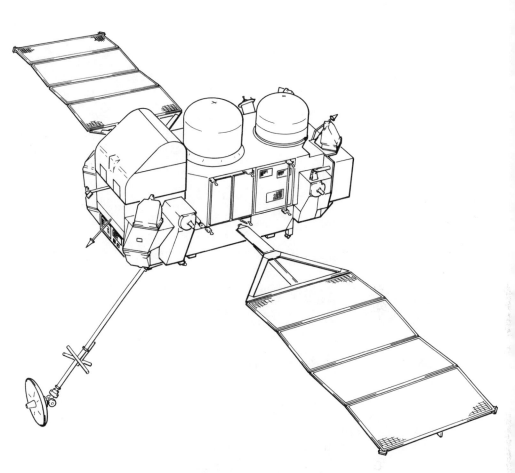

A schematic of the Gamma Ray Observatory. (*Courtesy of TRW.*)

U.S. and Dutch technicians prepare the Infrared Astronomical Satellite for launch. (*Courtesy of JPL/NASA.*)

The four-story Space Telescope is assembled in one of the largest clean rooms in the world at Lockheed Missiles and Space Company. (*Courtesy of Lockheed Missiles and Space Company.*)

_FOUR

Dewars and Thinkers:
The Infrared
Astronomical Satellite

16

In the early 1960s astronomy was in revolutionary ferment. New telescopes that could observe wavelengths outside the visible range were leading to new discoveries and a bold new spirit of adventure. The old order was giving way, not so much to a new order, because no one knew quite how to organize the new knowledge, but to a new perspective. The radio astronomers had led the way with their discovery of exploding stars, exploding galaxies, and the mysterious quasi-stellar radio sources—quasars, as they eventually came to be called. By 1961, David Heeschen was at the National Radio Astronomy Observatory in Green Bank, West Virginia, making plans to adapt Martin Ryle's ideas on linking many telescopes together to make a super radio telescope, plans which would come to fruition seventeen years later in the Very Large Array. About that time, Riccardo Giacconi was working for American Science and Engineering in Cambridge, Massachusetts, in a converted milk truck garage, putting together the rocket payload that would discover the first x-ray star in 1962 and thinking about an x-ray telescope, an idea that became a reality seventeen years later in the *Einstein* x-ray observatory. Carl Fichtel, at Goddard, was building a detector for cosmic gamma rays.

Frank Low, meanwhile, was working at Texas Instruments and had just completed a device that would open up yet another region of the electromagnetic spectrum, the infrared. The infrared is the name given to those frequencies of radiation ranging from just below the red portion of the visible spectrum down to radiowave frequencies. Objects having temperatures in a range from about 2,000° Celsius down to about 20° above absolute zero (−253° Celsius) radiate most of their energy at infrared frequencies. This includes such diverse objects as cool stars, people, planets, and grains of dust in interstellar space. Infrared radiation is sometimes called heat radiation because objects that are not hot enough to glow, yet give off heat, such as wood-burning stoves, are strong emitters of infrared radiation. Heat radiation is a misleading name, however, because all forms of electromagnetic radiation carry energy that can be converted to heat, as anyone who has seen a microwave oven in operation or suffered a sunburn from ultraviolet light can testify.

Near-infrared radiation, that is, radiation at frequencies just below the red, had been detected from the moon in 1856 and from other objects such as the planets and bright stars in the 1920s and 1930s. In the 1950s and early 1960s, a few astronomers, most notably Harold Johnson of the University of Arizona, tried to push infrared observations to lower frequencies, but these efforts were hampered by inefficient detectors—less efficient, in fact, than the infrared sensors used by snakes to locate mice or other warm-blooded prey at night.

The detector problem was solved not by an astronomer at one of the major observatories but by a physicist who was working on an altogether different problem. Frank Low had been interested in astronomy as a graduate student at Rice University, but only as an avocation. "I built one of these cheap 3-inch telescopes," he told us. "Houston was a terrible place to look at anything and I knew that . . . But I thought, well, I'd spend my sixty bucks or whatever it was that I scraped together and put it together myself and then I'd at least know what the elements of a telescope were. If we had had any kind of astronomy course or space-science courses at Rice at the time, I guess I would have taken them . . . They had none." Ironically, Low would return to Rice University several years later as a noted astronomer to become a professor in the newly created Space Sciences Department. But in 1959, when he finished graduate work, he was on his way to becoming what he had wanted to be since childhood, a physicist.

Low was born on November 12, 1933, in Mobile, Alabama. He was two years old when his father, Francis McFadden, died. Several years later his mother married Albert Low, a civil engineer. Frank Low's interest in science developed early. "Certainly as early as ten years old, I started reading all I could find that dealt with things like relativity," he recalled as we sat in his office on the University of Arizona campus. It was not an office calculated to impress—no carpets or paintings or couches or plants. Just stacks of papers and documents, not particularly well organized, shelves full of books, and an artist's rendering of the Infrared Astronomical Satellite. "Einstein was literally kind of a role model," he continued, then paused and chuckled. "Obviously that's too specific. I wasn't dreaming of being an Einstein. He was a hero in the same way that Roger Staubach was to young kids. They knew they weren't going to be a great quarterback, but they could relate to that kind of thing." Low attended high school at St. John's in Houston, one of the best private schools in the southwest. "I started getting into science right away, and each year science looked more attractive to me." Upon graduation, he won the math prize, but not the physics prize, which disappointed him. "I was really quite upset," he remembered. "I thought I should have gotten the physics prize and they should have given the math prize to someone else."

After high school he attended Yale. A strong interest in music paralleled and competed with Low's interest in science during his high school and college years. Encouraged by his mother, he became a dramatic tenor. "I debated all the way up through college whether I wanted a musical career or not." He sang in the Yale Glee Club and in light opera productions. "Fortunately," he said, "at the last possible moment I became persuaded against that, because I think I would have made a lousy musician." He obtained a bachelor's degree in physics from Yale in 1955 and returned to Houston to do graduate work in solid-state physics.

After obtaining his Ph.D, Low went to work at Texas Instruments near Dallas. "I was working on thin film superconductors," he recalled. Superconductivity is the flow of electrical current without resistance. It is a phenomenon that occurs in many materials; below a certain transition temperature, which is a characteristic of the material, electrical currents flow without resistance or loss of energy. This phenomenon has immense practical potential which is only beginning to be realized. The principal difficulty has been that the superconducting transition occurs at very low temperatures, within 20° of absolute zero, so the superconducting state can be destroyed by very small heat leaks into the system. Low's supervisor at Texas Instruments was King Walters. He suggested that Low investigate the possibility of taking advantage of the breakdown of superconductivity at the transition temperature to make a very sensitive energy-measuring device, or bolometer. The procedure, basically, is this: Cool a thin strip of material such as aluminum to just below the transition temperature, enclose it in a can of some sort that has a small opening, and measure the conductivity of the strip. If a small amount of energy, say in the form of infrared radiation, passes through the hole and strikes the aluminum, it will destroy its superconductivity. This radical change in the measured conductivity of the aluminum strip can be related back to a measurement of the amount of energy that struck the strip.

"There had been one or two papers in the literature on superconducting transition as a temperature sensor," Low recalled. "King Walters, my boss, suggested that I read these papers and see if there was any relationship, any possible tie-in there" with the research they were doing. Low quickly realized that the temperature sensors would not work as proposed. He wondered if a similar device that uses semiconductors rather than superconductors would work. "I worked for a company that was heavily into semiconductor research," he explained, "so it was a natural thing to see if semiconductors would solve the problem. It turns out it does solve the problem." Low immediately began to construct a bolometer that used germanium as its working element. This material can be prepared so that it will change its electrical conductivity when it absorbs energy. This change can be directly related to the amount of energy absorbed. The general principle of the semiconductor bolometer is the same as that of the proposed superconductor bolometer, but it works better. "The whole effort to develop the bolometer was a couple of weeks and I just did it without permission. I had an understanding boss, and I guess I must have told him I was going to take a little bit of time and see how this device worked. I knew it would work; I just didn't know how well it would work. It worked very well."

In the months that followed, he experimented further with the germanium bolometer, worked out the theory, and wrote a paper describing it. The publication of his paper in the *Journal of the Optical Society of America* in late 1961 changed the direction of Low's career. "When the paper came out, it was noticed by a number of people," he recalled. "That opened the door to my becoming directly involved [in astronomy]. What happened was that Frank Drake and Carl Sagan, who was just a young whippersnapper at the time, were very interested."

Both Sagan and Drake visited Low's laboratory at Texas Instruments to learn more about his new instrument. "Carl's thing at the moment was that if we had a good enough infrared detector we could probably beat the Russians to the discovery of life on Mars." In the late 1950s William Sinton of the University of Hawaii had made near infrared observations of Mars. He found three features in the infrared spectrum that he speculated could be due to the presence of organic molecules, and he published several papers on evidence for vegetation on Mars.

Drake, a radio astronomer who, like Sagan, has popularized the importance of searching for extraterrestial intelligence, was undoubtedly interested in the possible use of infrared detectors as an aid to the search for life in the universe. But he also had a more immediate and practical application for the bolometer in mind. "Frank was interested in using it for millimeter wave observations. He knew enough about it, having read the paper, to know that a very sensitive bolometer could be the answer to pushing radio astronomy into the millimeter regime." Previously, radio receivers had been sensitive only to radiation having wavelengths greater than about 10 millimeters (1 centimeter).

During this time a mutual friend arranged for Low to get together with Harold Johnson, who was at the time probably the foremost infrared astronomer in the world. Johnson was then at the University of Texas in Austin. Recalls Low, "I hopped on the plane and went down to UT and met Harold Johnson, showed him my paper, and he was very excited about it." The interest of astronomers in his work encouraged Low to continue developing his bolometer with an eye toward using it for astronomical observations. He soon made another important breakthrough.

"While I was still at TI I built this Dewar," he explained. A Dewar (pronounced "doer") is an insulated flask used to store liquefied gases at temperatures above −200° Celsius. It had a double wall with evacuated space between the walls and silvered surfaces to minimize heat loss. It was invented in 1892 by James Dewar, a Scottish physicist. The Thermos bottle is a familiar example of a Dewar flask. Low's bolometer worked best if it was kept at temperatures close to absolute zero. This required surrounding it with a bath of liquid helium at less than 4° above absolute zero. The existing glass Dewars were too delicate for the rigors of the observatory, especially if the observatory was to be in a balloon or on an airplane or a satellite; because they were inefficient, an astronomer would have to keep adding liquid helium, which is expensive. Low came up with a hardy, efficient metal Dewar. "The Dewar is almost as important as the bolometer," Low said. "The bolometer was the really important thing I did, but the Dewar was necessary to make these things viable. Today, twenty years later, almost identical metal

Dewars are everywhere." With his background in industrial research, Low was well aware of the commercial possibilities for his inventions. He patented them and in 1968 formed Infrared Laboratories, Inc., which is today a thriving company, with a dozen or so employees, that makes Dewars, bolometers, and other instruments for infrared and low-temperature research.

Drake, meanwhile, had been impressed with Low's work and its potential importance for astronomy. Upon returning to the National Radio Astronomy Observatory, where he was working at the time, he persuaded the director, David Heeschen, to make Low an attractive job offer. Besides his salary, he would have $50–70,000 a year in research funds to spend on the development of his bolometer.

"I had a tough decision to make there," Low recalled, "because I was going along pretty well at TI." He had been promoted twice in two and a half years, had earned a stock option, and was on the vertical ladder, headed for a senior research position or possibly upper management, a satisfying and lucrative prospect. Low decided to give it up for the spartan, splendid isolation of Green Bank.

"It was, in hindsight, a wise choice. I had no background in astronomy," he explained. "At Green Bank at the time, there was this very, very good group of radio astronomers—Heeschen, Drake, Wade, Dave Hogg, and others. All these guys were there, and there was this beautiful library and this isolation, so there wasn't much to do but to spend one's evening talking to the other astronomers or reading in the library or working in the lab. It was like going back to graduate school in a sense. I had the leisure. I didn't feel the pressure to go on to something instantly. I had time, which is a very valuable resource."

"From the point of view of being able to do valuable science, it was a lucky thing," he continued. "I said 'wise' earlier, but I think lucky is better. I lucked into it. Partly it was the efforts of people like Frank Drake that made it possible. I owe Frank a great debt and he knows it. I've said many times that it probably wouldn't have happened without his efforts."

Drake's motives were altruistic. He wanted someone to explore that part of the electromagnetic spectrum between radio and optical frequencies, and he believed that Low had the knowledge and the temperament to fill such a pioneering role. "I was given then what was fifty thousand, or maybe it was seventy thousand, dollars to spend each year that I was there," Low recalled. In today's dollars, that would amount to roughly $200,000 a year. "Imagine a young guy today being given a sort of free rein and $200,000 a year to go and do what he wants to do in the basement. That doesn't happen very often," he laughed.

It was a good investment. David Allen, an infrared astronomer at the Anglo-Australian Observatory in New South Wales, Australia, wrote in a review of infrared astronomy in 1977: "Now, 15 years later, if I were asked to name the man who has done the most for infrared astronomy, it would still be Frank Low, who, having invented the germanium-gallium bolometer, forsook his physics laboratory and turned his attention to pioneering astronomical observations with them."

One of the first astronomical observations made with Low's bolometer was of Mars, as Sagan had wanted. A bolometer system, complete with Dewar for a small telescope, ascended by balloon to 78,000 feet in 1963 and made infrared observations of Mars. The balloon was used to get the telescope as far as possible above the atmospheric water vapor and carbon dioxide, which absorbs infrared radiation. "A crude spectrum of Mars was obtained with it," Low recalled, "and although technically it was a worthwhile step forward, I think that it was realized that the features of the spectrum didn't have anything to do with life on Mars."

While at Green Bank, Low designed a 36-foot telescope for millimeter wave observations. Since much of the atmospheric absorption at one millimeter is due to water vapor, a high, dry site is essential. Kitt Peak in Arizona was selected, mainly because the site was already under development for a complex of optical telescopes to be operated by the Kitt Peak National Observatory. Low and his family moved to Tucson, where he worked for the National Radio Astronomy Observatory while the millimeter telescope was constructed. When it came time to return to the National Radio Astronomy Observatory headquarters in Charlottesville, Virginia, Low decided to stay. "Edith [whom he had married in 1957] and I decided that we liked it at Tucson; they gave me a job here and I've been here ever since."

In Tucson, Low began a collaboration with Harold Johnson, who had moved to the University of Arizona from Texas. Using Low's newly developed bolometer, they ventured into the unexplored region of the middle infrared, at wavelengths of .01 millimeter, or 10 microns. The first dragon that a would-be astronomer must slay if he is to survive in the middle infrared is that of the infrared background.

Under ideal conditions, the Low bolometer is so sensitive that, if it were attached to the 200-inch reflector on Mt. Palomar, it could detect the infrared radiation from a mouse as far away as the moon. Unfortunately, conditions are far from ideal. Almost everything on earth emits radiation in the near infrared—people, bugs, birds, the telescope itself, the building, the earth, even the atmosphere. There is no such thing as a clear, dark night for the infrared astronomer; the sky is always aglow with infrared radiation. These spurious sources of radiation must be eliminated or else an infrared observation will yield an incoherent blur. Telescopes intended especially for infrared observation are designed so that contamination by stray light from the struts or supports is reduced, and the detectors are immersed in a Dewar of liquid helium. However, most of the infrared background cannot be screened out. It must be coped with.

This is done by what is called "nodding." The telescope is pointed first toward the desired target, which is observed for a certain length of time. During this time, the telescope is picking up radiation from both the target and the infrared background. After a specified time, the telescope is pointed to a nearby "blank field," that is, a region of sky in which there are no bright infrared sources. During this time, the telescope is picking up radiation only from the background. By subtracting the results of the second observation from those of the first, the background can be eliminated. This technique, called sky subtraction, is used in every field of astronomy. In the infrared, though, the background radiation can be 100,000 times as strong as the radiation from the target, so a small variation in the background radiation would make the results for the target totally unreliable. Therefore, the telescope must continually monitor the background by looking back and forth from the target to the sky, or nodding, as rapidly as possible. In practice, this is usually accomplished by nodding a secondary mirror back and forth rapidly but very slightly, so that the image produced by the primary mirror is alternately on and off the detector. (In reflecting telescopes of the type used in major observatories, the incoming radiation is reflected first off the primary mirror, then off a secondary mirror, which is used to divert the radiation so that an image is formed at a more convenient location, outside the main body of the telescope.)

What Low and his colleagues found is that the universe is a dirty place. Old stars, young stars, stars still in the process of formation, all have clouds of dust around them. So does the center of our galaxy and the centers of many other galaxies, and quasars as well. Dust is everywhere. For his accomplishments, Low was awarded the American Astronomical Society's Helen B. Warner prize for significant contributions by astronomers less than 36 years old, a prize awarded to Giacconi two years earlier for his work in x-ray astronomy.

Thinking back on how he got into astronomy, Low offered this suggestion. "My advice to a young person is to be flexible, look at all the opportunities, and try to make the best choice that you can. Do the things that you know you can do. Or at least that you think you can do. You don't know. You have to guess at what you can do. Fortunately, I had some good opportunities all along."

In 1966 Low began dividing his time between the University of Arizona and Rice University, where he became a professor in the Space Science Department, partly at the urging of his former boss, King Walters, who had himself become a professor at Rice. Drawing on his own experiences, Low believes strongly that the best preparation for a career in astronomy is a solid diet of physics courses, and he was a persistent and vocal advocate of merging the Space Science and Physics departments at Rice. But his efforts were not nearly enough to budge the entrenched establishment, and the departments remained separated. Similar efforts at the University of Arizona have failed, to his dismay. "The physics and astronomy departments should be connected," he told us. "I think that's a great shortcoming at the University of Arizona, and at Rice." Low left Rice in the early 1970s after a productive period in which he and his students made a number of important discoveries.

It was during this time that Low led the way into the far infrared region of the spectrum. At wavelengths longer than about 30 microns (0.03 millimeters), the absorption of infrared radiation by the atmosphere is overwhelming, and infrared astronomy from a ground-based telescope is impossible. Undaunted, Low took to the air in a converted Lear jet and became possibly the first astronomer to make observations while wearing an oxygen mask. He and his colleagues found that the center of our galaxy and the centers of other galaxies were strong sources of far infrared radiation, as were clouds of gas and dust in our galaxy.

One of the most discussed objects discovered by Low and his

colleagues is the Kleinmann-Low Nebula, which he and Douglas Kleinmann, then a graduate student of Low's at Rice, found in 1967. The Kleinmann-Low Nebula is in the constellation of Orion and is part of a giant cloud of gas and dust. It has a mass of 200 suns and radiates as much energy as 100,000 suns; yet it is invisible at optical wavelengths. The enormous power of this infrared nebula can be supplied only by a massive star or by a group of stars embedded in it. It seems likely that most of the power is generated by a very young star that is less than a few thousand years old and has the mass of about 50 suns. Detailed studies at infrared and radio wavelengths indicate that the Kleinmann-Low Nebula contains a number of very young stars and clouds of gas that may be just now collapsing to form new stars. It may be the nearest interstellar "nursery" where we can observe the energetic and sometimes violent birth and development of stars.

The massive young stars or protostars responsible for the prodigious output of the Kleinmann-Low Nebula were first discovered and cataloged by Gerry Neugebauer and his colleagues at Cal Tech. Today, Neugebauer is a professor of physics at Cal Tech and director of the Palomar Observatory. The blackboard in his bright corner office is full of diagrams and equations. Snapshots of his family and photographs relating to the Infrared Astronomical Satellite mission share those spaces on the wall not taken up by bookshelves. Organized stacks of scientific reports and documents occupy the table and desk tops. His personal manner is in keeping with his office, pleasant and informal.

Neugebauer, like Low, came to infrared astronomy by way of physics. Also like Low, he cannot remember a time when he was not interested in science. "I always wanted to be a scientist, I guess," he said. "It was my father; that's basically the answer." Neugebauer's father is Otto Neugebauer, a distinguished historian of the sciences in antiquity. "He is a fascinating character," Neugebauer continued. "He's fun to talk to and he likes science and I guess I have just always," he paused to smile at the recollection of those early conversations with his father, "I mean, there's never been any doubt in my mind I was going to go on and do some form of science."

Neugebauer grew up around Brown University in Providence, Rhode Island, where his father is a professor. After high school, he attended Cornell University and in 1954 received a bachelor's degree in physics. At Cornell in those days, every male who had had not otherwise satisfied his military obligation had to train for the Reserve Officer's Training Corps (ROTC) and to report for duty upon graduation. Neugebauer asked for and received a deferment so he could attend graduate school at Cal Tech. In a very direct way, Neugebauer told us, the Army helped him to finish graduate school. After he had been at Cal Tech for a year and a half, "they wrote a letter asking when I was going to report for duty," he recalled. "And I said, well, the average time for graduate school is five and a half years. I got my orders four years in advance, to report to active duty in February 1960. I knew I had to get out," he laughed. "The pressure was very good for me."

When he did report to active duty, he was sent about ten miles away, to NASA's Jet Propulsion Laboratory. It was there that he was transformed from a high-energy physicist to an infrared astronomer, not as the result of any plan, but by accident. "I just happened to be in the right place at the right time," he said. "Rich Davies came to me and said we need to put an infrared experiment on a *Mariner* spacecraft that was going to Venus. So do it. So I did it. That's how I got involved in the space program. From the space program it was natural to come here."

Because the Jet Propulsion Laboratory is operated for NASA by Cal Tech, there are close ties between the two organizations and many professors have offices at both places. This brought Neugebauer into contact with Robert Leighton, a professor of physics with whom he had worked as a graduate student. "I started working with him in cloud chambers [devices used in high-energy physics to detect charged subatomic particles]. Afterwards he worked up at JPL on *Mariner 4* to Mars and I was peripherally involved in that. We talked about things, and he said he wanted to build a dish [an infrared telescope] . . . So we started doing that and it was just completely natural."

By 1960 it was well known to astronomers who had made infrared observations that the atmosphere is especially transparent at a wavelength of 2.2 microns, or about three times the wavelength of red light. But until Leighton and Neugebauer came along, no one had made a detailed survey of the sky in the infrared. "We decided that the right thing to do was a survey. I think he's the one that had the basic idea which has guided all of this . . . that if you make an unbiased survey you are going to learn just a hell of a lot." Leighton knew that it would be difficult if not impossible to get the time on a major optical telescope that would be necessary to carry out a comprehensive survey. His solution was disarmingly straightforward: we'll build our own telescope. Such an avenue was open because the longer wavelengths of infrared radiation allow for a somewhat greater margin of error in constructing and grinding the mirror than is demanded in the construction of high-quality mirrors used for optical observations.

Under Leighton's direction, a metal dish 1.5 meters (60 inches) across was machined to the approximate shape desired and then coated with epoxy and spun on its axis. Because of the centripetal acceleration, the epoxy assumed the shape of a parabola of revolution, just the type of surface needed to make a focusing mirror. (This is similar to the way coffee climbs up the sides of the cup when it is stirred.) Leighton kept the mirror spinning for three days, so that the epoxy would harden into the parabolic shape. The mirror was then transported to the top of Mt. Wilson where, for the next six years, with Neugebauer taking the lead, they made an infrared map of all the sky that is visible from Mt. Wilson, or about three-fourths of the total sky.

They detected over 10,000 infrared sources. But how many were real and how many were spurious sightings caused by airplanes, weather balloons, satellites, birds, or some other stray source of energy that could trigger the heat-sensitive detector? Neugebauer was determined that their infrared map would show no spurious sources, nor would any legitimate sources be thrown out with the spurious ones. As a result, over half the sources detected were rejected. The resulting catalog contained 5,612 sources. No entry has turned out to be spurious and only a few omissions are known.

Only a small fraction of the 6,000 or so stars visible to the naked eye are in the Cal Tech infrared catalog. The visually brightest objects have temperatures ranging from 4,000 to 20,000° Celsius (the surface temperature of the sun is 5,700°C), whereas the brightest objects in the near infrared (2 microns) have temperatures ranging from 1,000 to 4,000° and radiate only a small part of their energy at visible wavelengths. The list of the top-ten brightest stars in the near infrared includes some familiar names: Betelgeuse in Orion, Antares in Scorpio, Arcturus in Bootes, and Aldeberan in Taurus, all red giant or supergiant stars. They have reached a stage late in their evolution in which they have a hot inner core surrounded by a distended outer envelope. The sun will become a red giant when it is about 8 billion years old, or about 3.5 billion years from now. Most of the sources in the Cal Tech infrared catalog are red giants that, though invisible to the naked eye, can be detected with large telescopes.

"If anybody had thought about it at all," Neugebauer said, "they would have been able to predict very accurately the results of the 2-micron survey. There are a few hundred objects that are significantly different . . . but basically what you're seeing at 2 microns is an extension from the optical."

So Neugebauer, like Frank Low, decided that if he wanted to find some really different beasts for the cosmic zoo, he would have to venture into the middle infrared. In the late sixties he and his colleagues, especially Eric Becklin, now of the University of Hawaii, established a friendly but competitive rivalry with Low as they explored the sky at wavelengths in the middle infrared. They discovered the Becklin-Neugebauer object, one of the bright young stars that powers the Kleinmann-Low Nebula; they showed that one of the-largest and most enigmatic stars in the galaxy, Eta Carina, is a prodigious source of infrared radiation; they made detailed studies of the infrared radiation from the center of our galaxy in both the near and middle infrared; and they looked beyond the galaxy at exploding galaxies and quasars. In this latter work, the ever-cautious Neugebauer suggested politely that Low might have overestimated the far-infrared radiation from the exploding galaxies and quasars. Low repeated his observations with a more sensitive detector. It turned out that, by and large, Neugebauer was right, and Low revised his numbers downward.

The early work by Neugebauer and Becklin on the center of the galaxy received an assist from Guido Munch, an astronomer at Cal Tech. "He made it possible and easy for us to use the telescopes without having to apply for time. He said, 'Come up and spend the beginning part of my run.' That's how we were measuring the galactic center. All on time borrowed from Guido." He smiled and continued, "I've been very lucky in that the right kind of people were available. Leighton is an overwhelming person. And Guido Munch, he was just great to help us like that."

Another scientist Neugebauer feels especially fortunate to have been involved with is his wife, Marcia, whom he met while they were both at Cornell. She is now a senior scientist at the Jet Propulsion Laboratory. We asked him how it worked out, having two scientists in the same family. "For me it has been ideal," he replied. "Again, I've just been extraordinarily lucky, in the sense that she has a good job in a related field. She understands that putting out this catalog has been a massive job, and she's understood the fact that I haven't spoken a civil word to anybody for many months . . . As a matter of policy I try very hard not to bring problems home. When things are good, you talk about them, but I try not to bring problems home. But she understands when I'm talking about things and she understands when I'm working. So I think in my particular case, it's ideal."

Meanwhile, back at work, it soon became clear to Neugebauer that the next logical step was to do an all-sky survey in the middle and far infrared. If a middle-infrared survey would be difficult from the ground, a far-infrared survey would be impossible. They had to think in terms of a satellite outfitted with an infrared telescope.

17

While Gerry Neugebauer and Frank Low and other infrared astronomers were thinking about how to put an infrared telescope on a satellite to get a better view of the heavens, military researchers were thinking about how to put an infrared telescope on a satellite to get a better view of the earth. A military installation might be so well camoflauged as to be invisible by day, but because of the excess heat given off in the form of infrared radiation, it would appear as a glowing ember to an infrared telescope. The knowledge that the military was spending vast sums of money to develop detectors that were unavailable to astronomers frustrated them in the developmental years.

David Allen, writing in 1977, voiced these frustrations. "Developments now occur in military laboratories whence they are inclined to leak only when some aging general has been persuaded that further developments have made them obsolete or that the enemy already has one. One of the most infuriating things to an infrared astronomer is to know that a detector far better than he has access to is orbiting the earth attached to a large optical telescope, looking downwards!"

The military telescopes did not look exclusively downwards, though. The military has always—at least since the time of Galileo, and much earlier if you count the astrologer-astronomers that were in the hire of Egyptian pharoahs and Chinese emperors for purposes of national defense—realized the importance of keeping in touch with the astronomical community. For one thing, some of the best and brightest researchers are to be found there; in addition, astronomers are forever pushing their equipment to the limit and searching for new and more sensitive methods of detection, which may be of interest to the military. Thus it was that Low received a contract to do infrared observations from the Air Force Cambridge Research Laboratory in Massachusetts, the same group that sponsored the rocket flight in which Riccardo Giacconi and his colleagues discovered the first x-ray star.

"I had built a small bolometer array with a 28-inch telescope," Low recalled, "and we were trying to do a 10-micron survey from the ground. Basically what we proved was that it was very, very hard to do. We found a few objects, but the return rate per unit effort was sort of at the limit beyond what anyone was rationally going to do. The amount of time it would take to cover the sky was like a lifetime."

The Air Force abandoned the idea of a survey from the ground in favor of a rocket program. Between April 1971 and December 1972, a 16.5-centimeter liquid-helium-cooled telescope was carried above the atmosphere on seven rocket flights launched from White Sands, New Mexico. Two more flights were launched from Woomera, Australia, in September 1974. This project, which was called project HISTAR, was carried out by Russell Walker and Stephan Price of Air Force Cambridge Research Laboratories. In all, they accumulated about 30 minutes of observing time from which they compiled a catalog of over 3,000 objects. Preliminary versions of this catalog were circulated among infrared astronomers as early as 1973.

"We started studying their results," Low remembered, "and we found a lot of problems with them." Low's group found that they could confirm only one in five of the new sources listed in the catalog. "But," Low added, "we also found that they had definitely discovered new objects." Especially prominent were dense clouds of gas and dust and highly evolved, dust-shrouded stars. "And," he continued, "they had done this with this little bitty telescope."

Low knew the time was right to propose an infrared satellite. One of the first people he contacted was Neugebauer. "Gerry was interested. We decided that the time was approaching when such a step should be made." Others had made the same decision. "The Cornell group, with Martin Harwit and Jim Houck and others were involved with Air Force rocket payloads. They were talking up a satellite very heavily."

They found a capable ally within NASA in Nancy Boggess, who eventually became the NASA program scientist for the Infrared Astronomical Satellite. She was instrumental in getting NASA headquarters to take the concept of an infrared satellite seriously. They started out thinking small. "It was directed toward a very small effort," Low recalled. "An Explorer satellite. [*Uhuru*, the first x-ray satellite, and *SAS-2*, the first gamma ray satellite, were both Explorer-class satellites.] Sort of a quick and dirty survey, because none of us really knew what was appropriate at the time."

They strengthened their team with the addition of four more distinguished infrared astronomers: Walker, who had moved to NASA's Ames Research Center, Harmut Aumann, a former student of Low's, who became the NASA project scientist at the Jet Propulsion Laboratory, Fred Gillett of Kitt Peak National Observatory in Tucson, and Thomas Soifer at the University of California, San Diego. "I was trying to put together the best team I could," Low explained, "both to win the proposal and to make sure that the project was going to go."

Their proposal was selected. "We wrote a crappy proposal compared to some of the other proposals," Soifer acknowledged candidly, "but we had those two guys [Low and Neugebauer] and we got the nod." In recognition of the high quality of the other proposals, NASA added Michael Hauser and John Mather from NASA/Goddard, Houck from Cornell, and Rainer Weiss of MIT to the team to study the design of an infrared satellite. "They [NASA] liked the general idea but they didn't particularly like all the things we proposed to do. We were going to cover the sky in three months . . . the whole thing would have been a very modest program. There were debates as to whether this was the best approach or not," Low said.

The time was the mid 1970s and the talk might well have gone down as no more than talk, if not for the Dutch. It was a troubled time for the NASA space-science program. Budgetary belt-tightening by the Nixon administration, inflation driven by rising oil prices in response to the Arab oil embargo, plus large cost overruns in the *Viking* mission to Mars had already forced drastic cuts in the High Energy Astronomical Observatory program the year before, and the development of the space shuttle was beginning to have a financial impact. The Dutch space agency, on the other hand, was looking for a project. They were coming off the successful flight of their first all-Dutch program, the Astronomical Netherlands Satellite, a small ultraviolet and x-ray observatory, and were eager for more. And they had money. The high oil prices that were squeezing the United States economy were a boon to the Dutch, who have Europe's major oil refineries.

"It turned out," Low told us, "that the Dutch had secretly, on their own, contacted Ball Brothers [an aerospace contractor in Boulder, Colorado] and had their own design. They had already chosen a name, the Infrared Astronomical Satellite. When they found out that NASA was interested, they went to NASA and said, 'Look fellows, we're already ahead of you here. Can we get together? They realized that it was going to be more expensive than they had originally planned and NASA decided that an international project would be better than a quick and dirty all-American project."

In 1975 NASA notified the Low–Neugebauer group that the satellite would involve a collaboration with the Dutch. "We were told that we should sit down with the Dutch and work out our differences," Low said. "It was something of a shotgun marriage, but a successful one. Now," he smiled, "it's touted as an outstanding example of a true success of an international project."

Shortly after the Dutch–United States collaboration was announced, the United Kingdom joined the program, which officially became known as the Infrared Astronomical Satellite program. It was agreed that the British, through the United Kingdom's Science and Engineering Research Council, would build the ground station and operations control center at Rutherford Appleton Laboratory in Chilton, south of Oxford; the Dutch, through the Netherlands Agency for Aerospace Programs, would build the spacecraft, including the solar panels for electric power, the onboard computers, and systems for controlling and pointing the telescope; the United States, through NASA, would build the telescope, the main survey detectors, and the system used to keep the detectors near absolute zero.

From a purely political point of view, the Dutch connection seems to have been essential. "I don't think it would have happened if it hadn't been that the Dutch were coming in looking for this kind of collaborative thing at the time," Soifer told us. The diplomatic aspects of an international collaboration also helped to give the program a certain stability, especially when they began having what Neugebauer characterized as "an infinite number of crises." "Politically, if we hadn't had the Dutch–British connection," Low said, "I don't think we would have launched it. If it had been an all-American project in that much trouble, they would have scrapped it, because it was painful to the [NASA] managers at the time. It wasn't a big thorn, but it was a thorn. And they would have taken the old snippers and . . ." he imitated pulling out a thorn. "The international participation made it very, very difficult to cancel it. And," he smiled mischievously, "we relied on that. We had to. We had no alternative. So, that was a twist to it."

The overall scientific responsibility for the satellite was vested in the Science Working Group, which was to have nine members from the United States and nine European members, with joint chairmen from the United States and Europe. The European team leader, or co-chairman, was Reinder Van Duinen of the Kapetyn Institute, who had been the science leader on the Astronomical Netherlands Satellite. Before the launch of the Infrared Astronomical Satellite, he was succeeded by Harm Habing of the University of Leiden. On the United States side, a possible power struggle for the leadership was averted when Low declared he was not interested in the job. This left the way open for Neugebauer, who promptly declared that he did not want it either. "But," Low recalled, "with a certain amount of gentle arm twisting he accepted the job. And all and all, though it's been a difficult job for him to do, I think one has to say that he's done a fine job as team leader."

Some of Neugebauer's difficulties stemmed from the Byzantine management structure. NASA scientific projects are managed through one of its research centers. The original planning for the Infrared Astronomical Satellite was done at Goddard Space Flight Center, and the scientific staff there was developing a considerable expertise in infrared astronomy; so Goddard seemed like the logical choice. But Ames Research Center in Moffett Field, California, had already been designated as the center for infrared astronomy. Planning was under way there for a large infrared telescope designed to be carried into space by the space shuttle, and the Infrared Astronomical Satellite seemed like the logical steppingstone toward the larger telescope. Therefore, Ames was given responsibility for the Infrared Astronomical Satellite—but not total responsibility. NASA, in a characteristic move, decided to divide the responsibilities between Ames and the Jet Propulsion Laboratory (JPL), which was given "overall responsibility," a term that was as ambiguous in practice as it sounds. The thinking was that Ames might not have the managerial expertise to handle such a large project. Therefore, the United States scientific team—the people that were going to use the telescope and knew how it had to be built and what it had to do—had to work things out and check them with the European team, with NASA headquarters, with NASA/Ames, and with JPL.

"The management structure was too nebulous," Neugebauer complained. "There was a real difficulty in making sure that the things that we thought were important would in fact get implemented."

The design that emerged from the planning stage was a telescope with a primary mirror 57 centimeters (22.5 inches) in diameter and an array of electronic detectors sensitive to infrared radiation. The telescope was housed in an insulating vessel (one of Low's Dewars) that contained 475 liters (127 gallons) of liquid helium. The liquid helium kept the barrel of the telescope and the primary and secondary mirrors at 10° Celsius above absolute zero. The detectors were kept even colder, about 2° Celcius above absolute zero. The principal instrument was an array of 62 semiconductor detectors that covered the major portion of the infrared spectrum, from 8 to 120 microns. Two other instruments made it possible to get a detailed infrared spectrum and detailed images for bright sources. Overall the satellite was 3.6 meters (11.8 feet) long, 2.2 meters (7.2 feet) in diameter, and weighed 1,076 kilograms (2,383 pounds), very roughly the size and weight of a Saab automobile.

"There were a series of difficulties," Neugebauer continued. He leaned back and put his feet on the table that serves as his desk, his relaxed manner unaffected by the narration of what must have been some of the most nerve-wracking, tension-generating moments of his career. "The first big one that I remember . . . was in the mirror polishing. That was being done at Perkin-Elmer. The day before it was supposed to be done they broke off a tool and it made a big gouge in the mirror. That's the first of these bad decisions. How much do you want to polish it to get rid of the gouge? That was one of the hard decisions to make. Nothing is comparing apples and apples. It's always comparing apples and oranges."

It's a familiar story in big astronomy projects. The scientists want the instruments to be as good as they can possibly be within reason; NASA agrees. The disagreement comes in deciding what is meant by "within reason." To the scientists, it means as good as the existing technology can make it, without breaking the bank. To NASA, it means within the specifications and within budget. The question of how much to polish the mirror also came up with the *Einstein* x-ray observatory; not surprisingly, the answer was essentially the same for the infrared mirrors as it was for the x-ray mirrors. "It was cut on the short side according to what I think that most of the science team wanted," Neugebauer told us. "But just so it met the specs, the official specs. Which was too bad because it was better than that before the gouge had happened." But Neugebauer had no complaints about the in-flight performance of the mirror. "It turned out that it was fine," he said.

The detector array also turned out to be fine, but only after a series of calamaties that resulted in the postponement of the launch, the shifting of the responsibility for the detectors from NASA/Ames to the Jet Propulsion Laboratory, and the narrow aversion of disaster for the entire project. The plan was to amplify the signals from the detectors with a special type of transistor called a MOSFET, which stands for "metal oxide semiconductor field effect transistor." "The virtue of the MOSFET," Neugebauer told us, "is that it operates cold. So you can put it right in the focal plane."

NASA subcontracted with Rockwell International to build the detector array, including the amplifiers. As the Rockwell engineers tested the array under the watchful eyes of the science team, the scientists did not like what they saw. "The testing really wasn't good," Neugebauer said.

Low was more explicit. "They [the MOSFETs] were not only unreliable but they were also unstable and noisy." He wanted to use a different type of transistor for amplification called a JFET, for junction field effect transistor. "We recognized here [at the University of Arizona] early on the deficiencies of MOSFETs and tried to pursue the JFETs."

"Frank came up with a JFET which was much more stable," Neugebauer said, "but the trouble was that the JFETs don't operate at liquid helium temperatures. They operate much warmer."

Low continued to work on the JFETs, however, trying to come up with a design that would work. He had lost all faith in the MOSFETs, because they reduced the sensitivity of the detectors. "We had lost our original goal of sensitivity with the MOSFETs," he said. "It was debatable whether the mission was worth flying at that level of sensitivity. And they were unstable, so that we lost the DC [direct current], the low-frequency stable measurements." This meant that the amplifiers could not work without distorting the signal. To use such a system was unthinkable to Low. "That was just not possible," he said. But NASA/Ames thought that Low's concerns were exaggerated and that a successful mission was still possible, so they accepted the detector array from Rockwell in the summer of 1980, with the understanding that it had to check out satisfactorily in the further testing that would occur at NASA/Ames.

What then transpired has been variously described as a tremendous setback or a great stroke of luck. Electonic textbooks warn that MOSFETs are very fragile, that they self-destruct easily under the action of static charges resulting from normal handling, and that extreme care should be exercised in handling them by using a grounded-tip soldering iron and guarding against static charges. Apparently somewhere along the line, someone was not sufficiently cautious. "Something happened to the detector array after it was delivered," Low said. "A four-million-dollar piece of hardware about this big," he held up a 3 x 5 inch index card folded in half, "with 62 detectors and 62 amplifiers was delivered to the government by the manufacturer and about a third of it didn't work."

The managers and engineers at NASA and Rockwell were dismayed, but the scientists could scarcely conceal their jubilation. The accidental destruction of the MOSFETs was "the best thing that could have happened," Neugebauer said. Low agreed.

"It was a tremendous piece of good luck," he said. "Because if those MOSFETs had not been destroyed . . . and who destroyed them, we were never able to determine. At least, if they determined it, it's a very well-guarded secret as to how those MOSFETs were destroyed. Whether it was negligence or just what it was, who was responsible, whether it was the people who unpacked it, or the people who shipped it, or in the shipping, I honestly cannot tell you what happened. And I honestly must say that I don't care what happened." He laughed the laugh of a gambler who has seen his number come up at the roulette table. "Although it was a tremendous setback to the project and it could have led to the project being cancelled right there," he continued, "that was the opportunity to rebuild the focal plane and get rid of those damned MOSFETs."

It was getting late, and a crash program would be required. But would it be done at NASA/Ames or the Jet Propulsion Laboratory? By now, in the understated words of Neugebauer, "there was a feeling between the two centers which was not very good . . . There were loads of fights going on." Personnel at JPL would complain about the progress on a particular item, would offer unsolicited advice, and would use their management authority to apply pressure on NASA/Ames personnel, who would complain of interference and harassment. JPL argued that the program to rebuild the detector array had to be carried out at their laboratory. There was not enough time, they maintained, to continue to divide the responsibility for this crucial task between JPL and NASA/Ames. NASA headquarters agreed.

They also agreed to rebuild it with JFETs instead of MOSFETs, using a design suggested by Low that would allow the JFETs to remain slightly warmer than the detectors, so they could operate properly. Gerald Smith, then the Infrared Astronomy Satellite project manager at the Jet Propulsion Laboratory, appointed a "tiger team" to rebuild the detector array. To make up a tiger team, Neugebauer explained, "JPL says okay they will get their best people and they can haul them off of any other project. Bob White headed up the tiger team. He was just superb, because he understood how to use the scientists and how to use the engineers. They worked—we all worked—around the clock. It was a very hard time."

"It took about a year and a half," Low recalled. "It took the best efforts of several very good engineers and managers at JPL, and the effort we had here. We developed that whole business, proved it out, and flew it," he said with pride. "As anyone will tell you, it's just amazing how stable the focal plane was. A lot of the good science that is coming out is directly due to that; so I unabashedly take credit for some share of having accomplished that."

Neugebauer agrees. "Frank came up with the design. First of all, he came up with the JFETs and then he came up with the design for putting the JFETs on low thermal struts." The struts allowed the JFETs to operate in a slightly warmer environment, out of the focal plane, where the temperature was kept at 2° above absolute zero.

Other, unrelated emergencies came up during the crash program to rebuild the detector array. First of all, Mike Hauser of NASA/Goddard came across a disturbing report on the in-flight performance of detectors similar to the ones to be flown on the Infrared Astronomy Satellite. The standard safety procedure used in many satellite missions is to shut off the detectors when the satellite is passing through a region high over the South Atlantic where the concentration of charged particles is especially intense. Collisions of these particles with the spacecraft can alter the performance of the detectors or even permanently damage them. Ordinarily, if they are shut off during the passage through the South Atlantic Anomaly, they will be affected slightly, but these effects quickly disappear and the detectors operate properly soon after they are turnd on again. The report

Hauser read indicated that the detectors were behaving erratically for many hours after passage through the radiation belt.

"Frank tested that on similar detectors back in Arizona," Neugebauer recalled. "And then we had a big radiation testing program here, to understand what was really happening." The tests indicated that a similar effect would indeed plague their detectors. This crisis was resolved, Neugebauer recalled, when "a kid working with Frank, Erick Young, came up with a plan." Young's plan was to increase the voltage dramatically and in essence sweep the detectors clean of all the lingering effects of the passage through the South Atlantic Anomaly. His solution worked, but it required changing all the circuits because there had been no provision for voltage increases of the magnitude needed. "That wasn't a big job to change," Neugebauer explained, "but, you know, changing anything in a spacecraft a few months before it's supposed to fly is always risky."

"Then," Neugebauer said, "there was the thing with the Dutch. We had to take the telescope back to Holland to fit with the spacecraft. The telescope was ready. It was delivered by the subcontractor, Ball Aerospace, in the spring of 1981. But the detectors that would go inside the telescope weren't ready. The logical step was to either wait until the detectors were ready and ship the telescope with the detectors in place to Holland, or to ship the spacecraft to the United States where it could be fitted with the telescope when the detectors were finished. However, political considerations dictated that the integration of the telescope and the spacecraft should be done in Holland, and that it should be done as soon as possible, because the Dutch had completed the spacecraft and were anxious to fit it with the telescope.

"So," Neugebauer recalled with a smile, "we put in—because we had no detectors—a mock up of four or five detectors that went back to Holland." In May of 1981 the telescope was flown to Holland, where it was received ceremoniously. By October of 1981, the telescope had been carefully fit into the spacecraft and had passed vibration tests and other tests to assure its space-worthiness. Then it was shipped back to the Jet Propulsion Laboratory, where the whole thing was taken apart again when the detector array was ready.

In the meantime, the scientists working on the dectectors had encountered yet another problem. "The 100-micron detectors were bad," Neugebauer recalled. "We could measure here, under the background conditions that we expected, that they would be spiky." In other words, they would not respond to infrared radiation in a predictable and reliable manner, so that the background could be reliably subtracted away to yield the signal from a faint star or gas cloud or galaxy. "Now the only reason we could measure that was because we had the JFETs by now. If we hadn't had the JFETs we wouldn't have known the detectors were bad."

"So," Neugebauer continued, "we went back to Washington— this is now very late in the whole program—and said we wanted to take all the 100-micron detectors out and replace the detectors. That was a big thing." He smiled at the memory of how it really went. "We had to go back to Washington and get permission to make the change," he said. But they knew that they could probably make the change in less time than it would take to get permission to make the change; so the JPL tiger team began rebuilding the faulty detectors, fifteen in all. "This guy Bob Frazier did it on a weekend in his garage. It took that kind of talent. Frazier rebuilt them all. I think that by the time we came back from Washington," he laughed, "the change had been made. So that one worked out very well. But it was a huge risk."

By the end of 1981 the rebuilding of the detector array was completed. They put it in place, at the focus of the telescope, and fit the telescope with detectors back into the spacecraft. The Infrared Astronomical Satellite was completed. All that remained was to cool the system down to a few degrees above absolute zero and begin the final testing. Very soon they encountered more problems. "There was a breakdown in the 25-micron-band detectors," Neugebauer recalled. "There was a mistake made so that there was breakdown across an insulator. There was a question as to whether we should fly with only half of the 25-micron detectors."

They had essentially three alternatives. The first was to fly the spacecraft without fixing the detectors. The scientists opposed this plan. The second was to separate the telescope from the spacecraft once more, take it out of its liquid helium bath, find the problem, and fix it. This plan was opposed because of the dangers inherent in warming up and recooling the telescope; the thermal stresses created by this process might cause still more problems. The third plan was to separate the telescope from the spacecraft and try to solve the problem from the outside, without warming the telescope.

Jim Houck, a science team member from Cornell University, came to the rescue—long distance. He was in England, visiting with the British science team when he heard about the problem. It occurred to him that a simple change in the circuit—reversing the voltage bias across the faulty detectors—should make them work again. The Jet Propulsion Laboratory team tried it, and, to everyone's relief and delight, it worked.

The problems continued right up to launch—problems in vibration testing, problems with a microprocessor (it failed), problems with the Dewar (evidence of a leak). This leak led to fears about what would happen when the cover protecting the mirror was ejected once the satellite was in orbit. "The leak meant there would be gas inside of the Dewar," Neugebauer explained, "because it had been sitting for many months. The fear was that when they ejected the cover, all this gas would go rushing out and condense onto the mirror. And there were very solid arguments that that was bound to happen . . . In the meantime the whole spacecraft was getting very old. The components were getting flaky. Gerry Smith, who was project manager, wanted to launch it because it was getting so old. If we tried to fix the leak we probably would do more damage." Launch was set for January 25, 1983, from Vandenberg Air Force Base in California.

We asked Neugebauer if there was ever a time when he thought they would be stumped by one or more problems and the project would be a failure. "Yeah," he replied quickly. "Just before launch another problem came up. They found a bad capacitor. It was in a gyro or somethng. But then," he said, some of the almost forgotten tension creeping back into his voice, "they found that it was generically bad. The whole batch. They looked at the x-rays of them and they found that some solder was put on badly and so they were growing whiskers, so they would break down eventually" because the metal whiskers would allow the capacitors to short circuit. When they studied the circuit diagrams, they realized to their horror that many of the capacitors were single-point failures. A single-point failure is just what it sounds like—if there is a failure at that single point, there is an overall failure. For obvious reasons, electrical circuits for spacecraft are designed with as few single-point failures as possible, and those components identified as single-point failure components are subjected to rigorous testing and quality control.

How, then, did a batch of sloppily soldered capacitors make it into the circuits?

"They hadn't been identified as single-point failures because capacitors don't ever go bad," Neugebauer explained, smiling ironically. "Then suddenly they had a generic problem. In fact the day before launch, when we had the final preship review, there was— here I can only say what I heard, which is that there were people in NASA who said. 'Don't ship. Don't launch it, because it is too big a risk.' But they got overruled in the highest level in NASA."

We asked Neugebauer what his opinion was at the time concerning whether to launch or not to launch. "I thought we should launch. I finally thought we should launch . . . I was just sick and tired of the whole damn thing." He laughed. "And the argument swayed me that it was getting to be old. An old spacecraft. And I knew that if we didn't launch we would take it apart and then break things. That was already beginning to happen. You do one thing and then somebody gets tired and," he shrugged his shoulders, "it's broken. And so, I finally felt we should launch. My biggest fear," he said emphatically, "was that it would go up there—I mean, to be honest, I sort of had dreams that I would, that after we had sold it so heavily— that it would go up there and work perfectly and then not find anything." He laughed. "It worried the hell out of me. Boy, we would really have had egg on our face if it had done that."

Low was not worried about that. "I guess I had more confidence that if we would get the thing up there and get the cover off and the satellite pointed that we would get some astronomy out of it. But I had no idea that we would get ten months of ultra-high-quality data. I just didn't have that feeling."

Low hedged his feeling by making a wager with Fred Gillett. If the lifetime of the mission, which would be determined by how long the Dewar held the liquid helium, was less than six months, Gillett would buy Low a case of Walker wine, the premium wine produced as an avocation by science-team member Russell Walker. If it was longer than six months, Gillett got the wine.

"After all this bad luck that we had during the development phase, right through the flight-testing phase, which is where you don't want to have bad luck," Low said, "our luck changed from what looked like an unending series of misfortunes to just blind good luck. The weather at launch was amazing. That winter there was nothing but storms that ripped apart the California coast, yet we got our one little six-hour period in between storms so we could launch the first time. If we had missed that it was an abomination," he continued. "There was this big pot of superfluid helium there on top of that rocket. Almost an unimaginable thing," he shook his head and laughed with relief. "We got it off within one millisecond of T zero. The Delta [rocket] had been having problems before, but it was the most perfect Delta launch that they had ever seen. The launch was terrific," and then once more, "that launch was just spectacular. They put it within one foot of the right orbit."

The Infrared Astronomical Satellite was launched into a 900-kilometer-high polar orbit at 6:17 p.m. Pacific Time on January 25, producing a beautiful visual display in southern California skies. Contact with it was established both through the NASA tracking stations and during its first pass over the Chilton, England, ground station. The operations control team at Chilton began the in-orbit checkout of the satellite. The control program in the onboard computer was not working properly. The satellite would respond to the commands for a while and then it would stop responding. Unless this "bug" could be found, the satellite was helpless. They could not safely eject the telescope cover, they could not do the in-orbit checkout, they could not maneuver the satellite or make observations. The cover was heating up rapidly; if a solution was not found within a week, the mission could fail.

The Dutch team that had designed the onboard computer software was called in. They soon traced the problem to the sun sensor. This device was designed to warn the control computer whenever the spacecraft was in an unsafe orientation—for example, when the telescope was pointed too close to the sun or earth. The problem was that the sun sensor was intermittently generating a false readout that indicated that the spacecraft was in an unsafe orientation, when in fact it was not. In response to this false read out, the control program would shut down the spacecraft. The Dutch software specialists modified the control program to filter out the erroneous sun sensor signals, and installed it by remote control from the Chilton ground station. The cover was ejected and the checkout was completed on schedule.

"They changed the software and they made it work," Low said with admiration. "They were clever enough to do it. It was really neat." From that point on, Low said, "the Dutch spacecraft worked beautifully, and the British did an admirable job of operating the spacecraft as well."

A combined Dutch–British team was formed as early as 1978 to develop a system for operating the satellite in a highly automated way. It took about 18 man-years to develop the software that generated and scheduled the 18,000 or so observations that the satellite performed during its ten-month lifetime, but in the end the effort paid off handsomely.

"It was a program-intensive, not a people-intensive, operation," Neugebauer explained. "It was a very economical operation, because it was run by very few people. The whole scheduling was done by computer. This is, as far as I know, certainly on JPL things, the first mission that was programmed by computer." On a typical mission such as the Mars *Viking* expedition, a scientist would put in a request to look at some object, a proposed scenario for pointing the instrument would be generated, and it would be put into the computer. The computer would check the scenario to see if it was consistent with constraints as to how quickly the instrument could be moved, whether or not it would be pointing too closely to the sun, and so forth. If the scenario was unacceptable, a new one would have to be generated. "On *IRAS*," he explained, "what you did was to put in a series of priorities. You put in a whole list of what you wanted to do and your order of priorities. The computer figured out how to schedule them. So there was nobody who figured out a daily sequence. The computer checked it." As a result, the scheduling of observations required far fewer people. "It wasn't a team of a hundred people. It was three guys [one per 8-hour shift, which was eventually reduced to one per 12-hour shift for much of the mission] who did this. For something that's going on 24 hours a day, that was very impressive. It was all done on the European side, and it worked very, very well."

The whole mission worked so well that Low lost his bet. "The helium lasted ten months instead of six months, so I lost a case of Walker wine," he laughed. "It was the happiest bet I ever lost."

"It was very much a team effort," he said, in summarizing the development of the satellite. "I don't think any one person or even any one of the elements of the team could have brought this thing off. It required too many ups and downs, and pitfalls. We very well could have launched a completely useless experiment. Or we could have failed to have launched anything at all. These were very real possibilities," he emphasized. "And some people still think even with the big success it has been that it was luck. Pure luck," he said slowly, with a certain mocking tone to indicate that he was not one of those people. Rather, he said "it was a tremendous technological event."

"What we were doing with *IRAS*," he explained, "was pushing several areas of technology, which either were very new or unknown to us. Infrared astronomy is so clearly related to infrared military programs, using basically the same kinds of technology for infrared sensing and infrared surveillance both on ground and in space. There is obviously a great overlap there. But I believe it's true that the *IRAS* satellite represented a culmination of technologies that had never been used in quite the same way that we used them. The long-lived helium-cooled cryogenic system was certainly one of them." The silicon detectors had been developed by the military. The germanium detectors, which made it possible to observe in the far infrared, had not been developed. "We were pretty much in virgin territory there," Low said. "One of the things that *IRAS* had to do was to quickly bring along the understanding of how to manufacture suitable germanium detectors. And we just barely succeeded."

"It has been a fantastic, very rich experience," he continued. "And NASA deserved a lot of credit for it, although at times we cursed NASA, because you get this adversary relationship between the NASA management and a science team. And the NASA engineers are under the thumb of the NASA management. They keep them that way. I think that most space projects have worked out in one form or another with the same sort of relationship. An adversary relationship. And you wonder whether that's good. It makes for ulcers, in some cases heart attacks [Jim Houck suffered a heart attack in early 1984; fortunately, he recovered and returned to work several months later], and bitter enemies. Although, I don't think we've got any bitter enemies now. I really don't. It's a stressful way to do a project. That's an element in it. Maybe the system we've got is the best that can be done. I just question that it is. It's tough on some people."

Neugebauer agreed. While saying that he did not mind big projects and was ready and willing to work on the next large infrared project, the Shuttle Infrared Telescope Facility, he added that "I don't think I'll ever be a project leader again." For one thing, he had had his fill of the 5,000-mile flight from Los Angeles to London. "I made twelve trips in twelve months. I got to be one of these people that know the stewardesses . . . It wasn't bad, but it was not very pleasant," he paused, then added, "I'm never getting on that flight again."

18

Twice a day the Infrared Astronomical Satellite transmitted 700 million bits of image data back to earth. Much of it was from irrelevant or bogus sightings. "The detector array generates 100,000 detections per day," said John Duxbury of the Jet Propulsion Laboratory, "and only 20 percent are real, stationary sources that you want to retain." The other 80 percent were signals produced by charged particles interacting with the detector, or stray moonlight, or infrared radiation from orbital trash, such as exploded boosters, burnt-out rocket stages, worn-out satellites, working satellites, clouds of particles ejected from solid-fuel rocket motors, nuts, bolts, flakes chipped off the satellite by micrometeorites. The telescope could detect a speck of dust at a distance of two kilometers.

"We have argued a long time to make a confirmation strategy that got rid of everything except rock-steady celestial objects that didn't move," Neugebauer explained. "The things they [the data reduction team] go through before a source is really accepted are horrendous."

Their basic strategy was to require that a source be confirmed by repeated observations over several time scales. If a source was seen by a row of detectors on one side of the detector array, and then by a row on the other side a few seconds later, they knew it had persisted for longer than a second. This would rule out spurious detections caused by charged particles. If a source was picked up on successive orbits in the same location, they knew it had not changed positions for about an hour and a half. They would then go back to the source about a week later and make two more scans; a few months later they would make yet two more scans of the source. By the end of the mission they had scanned 72 percent of the sky six times, and 95 percent of the sky four times. "The important thing," Tom Soifer, the member of the science team who oversees the data processing emphasized, "is to produce a very high-quality catalog." Their goal was to have less than 2 false sources out of every 1,000, and to throw out no more than 2 of every 100 real sources. In all, about 250,000 celestial sources of infrared emission have been cataloged. "We had an overwhelming emphasis on reliability," he continued. "We had to see things many, many times to be convinced of their reality. And it made the data processing for that reason rather complicated." Rather indeed. They have to keep track of about a million detections, which must be checked and cross-checked to establish their reality. According to Duxbury, it has taken more than 100 man-years to develop the data-processing facility.

It is possible that an object could fail the rigorous criteria required to make the catalog and still be of considerable astrophysical importance. Examples would be a possible tenth planet or a faint companion star of the sun, since these objects would change position from one detection to the next. With this in mind, the science team is building up a file of rejects. "We are going to have a file of rejects," Neugebauer said, "and they're going to be fine pickings for people who look for curious events."

While the United States team was developing the procedures for constructing the catalog, the British team developed procedures for studying one important class of rejects. They realized that the requirement that sources be detected on successive orbits might rule out some important fast-moving objects, such as asteroids or comets. They modified their program to allow the data to be searched for such fast movers. This effort paid off with the first dramatic discovery of the mission.

Less than a month after launch, the program to detect fast movers had identified a number of known main-belt asteroids. Main-belt asteroids are rocky objects in roughly circular orbits around the sun between the orbits of Mars and Jupiter. The procedure was to send an "alert" to a cooperating observatory where the object could be studied optically and usually identified. On April 26, 1983, three months after launch, the Infrared Astronomical Satellite detected a fast mover and sent out an alert. Two days later, observers in Sweden photographed it through earth-based telescopes. Its fuzzy appearance indicated that it was a comet. On May 3, two amatuer astronomers, Genichi Araki of Japan and George Alcock of Great Britain—both unaware of the satellite discovery—independently found the comet. Following the standard procedure for recording the discovery of comets, the amateur astronomers reported their find to the Central Bureau for Astronomical Telegrams in Cambridge, Massachusetts. The comet was officially named IRAS-Araki-Alcock. It was the fifth comet to be named for the 71-year-old Alcock. A legend among amateur and professional astronomers alike, Alcock, a retired English school-master, has discovered more novae (four) than any other astronomer, and he undoubtedly knows the positions of the stars better than anyone else who has ever lived. He has memorized the positions of 30,000 stars, roughly five times as many as are visible with the naked eye.

IRAS-Araki-Alcock was unusual because it passed within 3 million kilometers of the earth, closer than any other comet in the past 200 years; so it was studied in detail both by the Infrared Astronomical Satellite and by ground-based observers. It moves into the inner solar system once every thousand years, swings around the sun, coming about as close to it as the earth does, then starts an outward journey which will take it far beyond the orbit of Pluto before it turns around and heads back toward the sun. Comets are thought to be mixtures of ice and rocky material that resemble dirty snowballs some 10 or 20 kilometers in diameter. As a comet moves from the cold reaches of the outer solar system into the inner solar system, the increasingly bright sunlight evaporates some of the ice and dust from the main part of the comet. Sunlight and particles flowing away from the sun sweep the ice and dust into a long, flowing tail.

The infrared observations of IRAS-Araki-Alcock revealed a wide, bright tail that extended for as much as 400,000 kilometers, or roughly the distance from the earth to the moon. The material in this tail, which was invisible to optical telescopes, must be dust, because ice would have been vaporized in traveling from the head to the tail. The infrared observations of the comet indicate that it was losing mass at a rate far faster than earlier estimates suggested. Even with the increased estimate of mass loss, IRAS-Araki-Alcock should last for several million more passages into the inner solar system.

In all, six comets were discovered by the Infrared Astronomical Satellite, and five known comets were also studied. In general, the infrared observations revealed more extensive tails than were visible from optical observatories. One comet, Tempel 2, was found to have a tail 30 million kilometers long. In fact, most comets probably have a trail of debris that stretches out along much of their multimillion-kilometer orbit. When the earth passes through this debris, the collision of the debris with the atmosphere produces spectacular meteor showers, or shooting stars. For example, twice each year, in May and October, meteor showers occur when the earth passes through the trail of debris from Halley's comet. Many of the well-known meteor showers have been identified with known comets, but not all of them. For example, until the flight of the Infrared Astronomical Satellite, no source had been identified for the Geminid meteor shower, which occurs in December.

But on October 11, 1983, a fast-moving object was detected which solved this mystery. This object, called 1983TB, has a sharp rather than a fuzzy image, so that it is not a comet. No evidence of cometary activity has been detected by either the satellite or ground-based observers. Rather, the object is an asteroid of an unusual type called an Apollo object. Apollo objects are rocky objects whose orbits cross the orbit of the earth. Approximately 80 such objects have been discovered, and it is estimated that there may be about ten times that many.

Where do they come from? The scientists who believe they come from the asteroids have a difficult time showing how they got out of their orbit in the asteroid belt to an earth-crossing orbit without being destroyed. Others think they are extinct comets whose ice and gas have been driven away over the course of many passages by the sun, leaving a rocky core. Until the discovery of 1983TB, they had little proof of this hypothesis.

Continued observation of 1983TB showed that its orbit corresponds very closely with the orbits of the Geminid meteoroids. Apparently 1983TB, which is now less than 2 kilometers across, is the parent body of these meteoroids. Since comets are the only objects that are known to be the source of meteor showers, it is natural to assume that 1983TB must have once been a comet. The highly elongated orbit, which is steeply inclined to the plane of the orbits of the planets, is also similar to the orbits of many comets. Why would 1983TB have become an extinct comet and not IRAS-Araki-Alcock or Halley's comet? Because it comes to within 15 million kilometers of the sun, closer than any known body in the solar system, ten times closer than the earth, and three times closer than Mercury. During these close passages, which occur every year and a half or so, it would have had its ice and volatile materials boiled away at a much more rapid rate than typical comets, leaving only a naked rock with millions of kilometers of debris trailing along behind it.

Iwan Williams of Queen Mary College in London has made the

unsettling suggestion that 1983TB may have a rendezvous with the earth in the not-too-distant future. According to his calculations, it is coming closer to earth each time it swings around the sun, and in the year 2115 it will pass between the earth and the moon, and may even hit the earth. Such an impact would release energy equivalent to 100,000 one-megaton bombs and would have catastrophic effects. Some scientists have suggested that mass extinctions of animals, such as the one that ended the age of the dinosaurs, have been caused by just such collisions.

Before we take to the street with sackcloth and ashes and signs announcing "The End is Nigh," we should bear in mind that, even if 1983TB comes as close to the earth as Williams' calculations indicate, the chances of it hitting the earth are still less than one in a thousand. Nevertheless, the probability is high enough, and the consequences grave enough, to keep astronomers from ignoring it altogether. A Spacewatch project has been funded by NASA and several private and corporate supporters; in this project, a specially designed camera is used in conjunction with a telescope on Kitt Peak to scan the sky for Apollo objects and keep track of their orbits.

Roughly six comets are discovered every year. Each one of these, as it nears the sun, will begin to form a long tail of dust. In the 4.5 billion years since the solar system was formed, billions of comets must have made their way into the inner solar system and left behind a trail of dust and gas. In addition, collisions between asteroids have surely added more dust to interplanetary space. Much of the interplanetary dust will have spiraled into the sun or accreted onto planets— the earth alone accretes some 100,000 tons per year—but some of it remains, concentrated in the zodiacal or ecliptic plane, that is, the plane in which the planets orbit. It is called the zodiacal dust. In the middle infrared, around 20 microns, the main contributor to the infrared background radiation is from the zodiacal dust. For this reason, one of the top priorities of the scientific team was to use the Infrared Astronomical Satellite to study the zodiacal emission in detail. They were surprised at what they found.

"I got a real thrill with the zodiacal emission, the zodiacal dust bands as they are called," Low told us, "because I guess I was the first one to realize what was taking place." What they found was the main zodiacal band, as expected, but also two outlying bands just above and below the main band. "Here, in a part of the solar system which we studied for years and years," Low explained, "only two astronomical units from the sun [an astronomical unit is the distance of the earth from the sun, 150 million kilometers], one astronomical unit away, was this absolutely peculiar structure in the sky which had never been seen optically! Although it could have been, it never was. And although it's maybe not the most fundamental aspect of the evolution of planetary systems, it's a kind of nifty thing and it does tell you something we wouldn't learn otherwise about the system."

The bands are thought to be caused by rings of dust particles that are tilted by about nine degrees with respect to the plane in which the planets orbit, like two wobbling hula hoops. What caused the rings? The most popular idea is that they were caused by the disintegration of a comet, perhaps because of a collision with an asteroid.

Whatever their explanation, the discovery of the zodiacal bands epitomized for Low the reason why he got involved in the project, and perhaps more fundamentally, in science. "The privilege of being able to just look at the data and the perception that this is the first time that anyone has ever seen this. And it's just lying there waiting for somebody to look. That is a big thrill."

Zodiacal dust and gas are a minor component of the solar system today, but once there was very little in the solar system other than the sun and dust and gas. The currently favored theory for the formation of the solar system goes something like this: An interstellar cloud of dust and gas, initially very extended and rotating very slowly, collapsed under its own gravitation: As it collapsed, the cloud began to rotate more rapidly, just as a figure skater increases her rate of rotation by pulling in her arms. This rotation caused the cloud to flatten into a disk. The collapse occurred most rapidly at the center of the cloud, producing a large clump in the disk. Eventually this central clump became hot and dense enough to be a star, which in the very early solar system was presumably surrounded by a disk of dust and gas. Over the course of about a hundred million years the dust grains in the disk began to stick together to form larger dust grains, which stuck together to form still larger dust grains, and so on until the planets were formed.

Early in the Infrared Astronomical Satellite mission, when the instruments were being checked by observations of bright, nearby stars, the scientists found a system that bears striking resemblance to this picture of the early solar system. They were observing Vega, a hot star that is about 2.5 times as massive as the sun, 50 times as luminous, and much younger, with an age of a few hundred million years. Vega had been observed extensively at wavelengths from the ultraviolet to the middle infrared, around 20 microns, so astronomers thought they knew it well. They were surprised to find that the infrared radiation beyond 20 microns did not fall off as suspected but remained almost constant. Further observations and theoretical arguments have convinced the science team that Vega is surrounded by a disk or ring of grains a millimeter or more in diameter. The mass in these grains is at least a million times greater than the mass in the zodiacal dust bands.

What is the origin of this extensive shell of dust and gas? Four alternatives have been considered. The first is that it is continuously produced by the star. There are many examples of hot stars surrounded by cooler matter that is expelled from the surface of these stars in stellar winds. But these stars all show clear evidence of this continuous outflow in their optical and ultraviolet spectra; Vega shows no such evidence. Another problem is that small dust grains of the type produced in stellar winds would have long since been blown away by the pressure of radiation from Vega. The effect of radiation pressure on grains also rules out the second possibility, namely, that the material was produced during an episode of mass loss that has long since passed. The third possibility, that the grains were accreted from the interstellar medium, can be eliminated by the same argument: the dust grains in the interstellar medium are too small. This leaves the possibility that the ring of dust is left over from the original cloud of dust and gas which formed Vega. In a paper discussing this discovery, Harmut Aumann and his colleagues concluded that "the shell around Alpha Lyrae [Vega] must represent an evolutionary stage intermediate between the final phases of star formation and the present state of our solar system."

Does this mean that planets have formed or will eventually form around Vega? This is, of course, the question that everyone would like to answer. It bears directly on the question of whether other star systems such as ours, with other planets such as the earth and other forms of intelligent life, exist in the galaxy. Unfortunately, the Infrared Astronomical Satellite cannot answer this question. Matter radiates more efficiently when it is spread out as in a cloud of dust than when it is clumped together, as in a planet, because of the larger surface area—that is why a baked potato cools more quickly when it is mashed and spread out over a plate. For this reason it is not possible to tell how much material is present around Vega in the form of planetesimals or planets. All that can be said is that there is an abundance of dust around Vega and that at least the first steps in the growth of interstellar dust grains to form planets has taken place.

Exactly one year after the publication of the discovery of a ring of dust around Vega, Low and colleagues Donald McCarthy, Jr., of the University of Arizona and Ronald Probst of Kitt Peak National Observatory published a paper describing evidence for the existence of an extrasolar planet. They used ground-based telescopes in Arizona to find evidence for a companion object orbiting a faint star called Van Biesbroeck 8. The companion, which they called Van Biesbroeck 8B, has a temperature of about 1,100° Celsius, an ideal temperature to show up at the near infrared wavelengths at which the group was observing. The inferred mass of Van Biesbroeck 8B, is somewhere between 30 and 80 times that of Jupiter. An object such as Jupiter would be too cool to detect. Van Biesbroeck 8B is too small to sustain nuclear fusion reactions in its core; yet is much more massive than Jupiter, the largest planet in our solar system. For this reason, some astronomers have questioned the group's claim that "these observations may constitute the first direct detection of an extrasolar planet." What the group may have found instead, they say, is a brown dwarf, an object that lies in limbo between stars and planets. All agree, though, that it was an important discovery.

In stars such as Vega and Van Biesbroeck 8, only a remnant of the dust and gas from which the stars formed has been captured in a dust ring or a planet or brown dwarf, and the infrared radiation from these systems is only a small fraction of the total radiation from the central star. In younger systems, in which the central star is still in the process of collapsing, or has just formed, the situation is reversed. These stars are still enveloped in the matrix of dust and gas from which they were formed. The radiation from the central stars, or protostars, is absorbed by these envelopes and reprocessed to infrared wavelengths; so infrared observations and, to a lesser extent, radio observations provide virtually the only means of studying such systems. The most massive young stars and protostars were well known from earlier infrared studies of objects such as the Kleinmann–Low Nebula. A major contribution of the Infrared Astronomical Satellite has been to extend these studies to stars more like the sun. In the years to come, as the data from the survey are sorted and analyzed, we should have a much more comprehensive picture of one of the most fundamental and puzzling questions in astronomy, namely, the question as to how stars form.

The discoveries that excite both Low and Neugebauer the most may be related to star formation on a grand scale. The science team has found evidence of starbursts in which millions of stars have formed in short bursts of a few million years. The vast amounts of infrared radiation generated by the starbursts produce infrared galaxies. "For me personally," Neugebauer told us, "the things that have intrigued me the most are the infrared galaxies. Those are the ones that I think are the most interesting, that will have the largest long-term impact." Low agreed. "The thing that's really interesting to me, too, is just the tremendous variety that's showing up now in the properties of galaxies. Infrared galaxies have been fascinating, the motivator, I think, of Gerry and myself. They kept us going." Nature has not disappointed them. "Galaxies are just turning out to be far more interesting and complicated than we knew," he continued. "*IRAS* has already detected 20,000 of them in the survey and many more can be dug out of the data."

Galaxies range from dust-free elliptical galaxies, which give off a negligible amount of infrared radiation, to our Milky Way Galaxy, which emits about half its energy in the infrared, on up to galaxies that emit more than 99 percent of their energy in the infrared. Arp 220 is the best-studied example of this new class of galaxies, the infrared galaxy. Optically, it is bright enough to be seen with a 10-inch telescope, and it has been known to astronomers at least since the turn of the century, when it was listed as number 4553 in J. L. E. Dreyer's *Index Catalog* of celestial objects. Over half a century later Halton Arp, the iconoclastic collector of weird galaxies, included it as number 220 in his *Atlas of Peculiar Galaxies*, along with the comment that it was a "galaxy with adjacent loops." The optical image of Arp 220 clearly shows these loops and, perhaps more importantly, two bright, oval-shaped smudges in the southwest corner. These bright ovals are thought to be galaxies in collision.

A collision between galaxies is not like a collision between billiard balls or freight trains, because galaxies are mostly empty space. It is more like a collision between two flocks of birds. Very few if any of the birds would actually collide, but many of them would undoubtedly alter their flight path out of fear or confusion, and some of them might even wind up in the wrong flock. When galaxies collide, very few if any of the stars suffer head-on collisions, but many may change their course and motion. Computer simulations of galactic collisions by Alar Toomre of MIT and Juri Toomre of the University of Colorado and more recently by K. D. Borne of the Carnegie Institution and J. G. Hoessel of the Space Telescope Science Institute have shown how they can produce bizarre configurations such as those observed in Arp 220 and other colliding galaxies.

The most spectacular effects are produced not by the stars but by collisions between clouds of dust and gas, or by the effects of the gravitational disturbances on these clouds. The collision of waves near the seashore give a rough illustration of what happens. As outgoing waves reflected from a rocky cliff encounter incoming waves, they can interfere with each other, producing a larger wave, perhaps even a spray of water. When galaxies collide, dust and gas clouds can be pulled out into graceful loops or bridges, or they can be compressed. Such compression might trigger a burst of star formation. The collapse of a cloud to form stars releases a tremendous amount of gravitational energy, especially when massive stars are formed. Thus, the formation of millions of stars over a short time could radically alter the luminosity and appearance of a galaxy. In recent years, observations from radio, infrared, optical, ultraviolet, and x-ray wavelengths have led to an increasing appreciation of the importance of this phenomenon in a wide variety of galaxies. Theorists, the speculators of astronomy, have of course taken notice, and starbursts now rival black holes as the proposed solution to many cosmic riddles—riddles such as Arp 220.

When the observations were scheduled for the Infrared Astronomical Satellite, Arp 220 had gained a reputation through radio and optical observations of being a weird object, but not weird enough to be singled out for special treatment. There was no evidence at the time, for example, that it might be undergoing a burst of star formation. It was observed in the course of the all-sky survey, and its infrared brightness at four different wavelengths was duly cataloged. Then in January of 1984, after the satellite had ceased to operate, Tom Soifer—who, like Low and Neugebauer, has had an abiding interest in galaxies—was going over a list of galaxies that showed up as bright infrared sources. What he found surprised and delighted him. Arp 220 was emitting about 80 times more energy in the far infrared than in the optical wavelength band. That meant that it was one of the most luminous galaxies ever discovered, comparable in brightness to quasars, those distant, enigmatic powerhouses that are thought to have been produced by large quantities of matter falling into supermassive black holes in the nuclei of galaxies. But quasars do not emit 99 percent of their energy in the far infrared, as does Arp 220.

"We hoped we would find this kind of stuff and, indeed, we are," he told us. "It's exciting." What is causing this incredible outpouring of energy? Could it be some peculiar sort of quasar, one embedded in dust, so that the outflowing radiant energy is degraded to infrared wavelengths? "I fluctuate almost on a daily basis," he admitted, "between calling this a dust-embedded quasar or a starburst. I'm now leaning toward a starburst. A week ago I was leaning toward a dust-embedded quasar," he laughed. "But the thing that surprised me is that, now, it appears that episodes of rapid star formation are able to produce as much luminosity as a quasar."

The infrared radiation is centered between the two optically bright ovals. Is this because that is where a burst of star formation is going on, or is it a bright nucleus buried in an envelope of dust? If the latter were so, then the double nature of the source might be an illusion. The dust band might simply be blocking the light from the middle of one large galaxy, splitting it in half so that it only appears to be two galaxies. Yet a survey by Carol Lonsdale of JPL, Neugebauer, and Soifer indicates that interacting galaxies are associated with a significant fraction of the extremely bright far-infrared galaxies found by the Infrared Astronomical Satellite. Low has suggested a hybrid model, wherein collisions between galaxies trigger a burst of star formation in the center of one of the galaxies; this, in turn, produces a bright quasar-like infrared galaxy.

Since the publication of the results on Arp 220, a team headed by Jim Houck has reported the discovery of six more infrared galaxies which have luminosities comparable to Arp 220. In Low's words, infrared galaxies are "a fascinating subject that we'll be dealing with for quite a while." Nor does he think it is by any means the last discovery to come out of the Infrared Astronomical Satellite. "I think what people will find is that it [the data file from the satellite] is really a treasure, a gold mine."

_FIVE

*Where the Stars
Don't Twinkle:
The Space Telescope*

_19

The greatest achievement of astronomy in this century has been a collective one—the opening up of virtually the entire electromagnetic spectrum, from radio waves to gamma rays. Out of this effort—which has involved hundreds of scientists, engineers, technicians, and not a few astute managers and politicians—has stemmed all the fantastic astronomical discoveries of the past twenty years: radio galaxies, quasars, x-ray stars, microwave background radiation, pulsars, black holes, gamma ray bursters, infrared galaxies. While we are justifiably awed by the brilliance of the discoveries made by the telescopes sensitive to radiation outside the visible band, we cannot ignore those telescopes that we have, over the years, come to know and love. Optical telescopes still have the starring role in the exploration of the universe. Most astronomers use optical telescopes more than the other types and will continue to do so for some time to come. One reason is tradition. Astronomy began with visible light observations, and such observations will always be the reference point against which we judge all others. Another is convenience. There are still many more optical telescopes than any other kind, and because they are on the ground and not in space, they are easier to get to and use. Finally, the optical wavelength band, like every other wavelength band, has its own intrinsic advantages. The radiation from the sun and from stars like the sun, of which there are billions in our galaxy alone, is primarily in the optical range. The same is true of other galaxies. Many important spectral lines, from which we can deduce the make-up and motions of stars and gas in space, fall in the optical band. These advantages are not overwhelming, and in the distant future we can expect astronomers' activities to be more or less equally divided among the various regions of the electromagnetic spectrum; they will have to be if we are to have a balanced perspective of the universe. But for the near future, observations in the optical band will remain the axis around which astronomy revolves.

Major new optical telescopes have been built in the last two decades in Arizona, Hawaii, Chile, Peru, Australia, and the Soviet Union. Supersensitive new solid-state detectors have enormously increased the capabilities of these instruments. About the size of a postage stamp, solid-state detectors are called charged coupled devices (CCDs). They produce small amounts of electrical charge in those areas of the detector that are struck by light. By measuring the charge that accumulates in each part of the detector, it is possible to get a very accurate estimate of the amount of light that hit it. They are over a hundred times more sensitive than the photographic plate techniques used in the 1950s and over three times more sensitive than the best image intensifier tubes available today. Charged coupled devices, when used with computerized image-processing techniques, can produce images far superior to what was possible only a few years ago.

Charged coupled devices detect up to 75 percent of the light that strikes them. Even if it were possible to make a perfect detector that would detect 100 percent of the light, astronomers would realize only a one third gain in efficiency. To achieve a much larger increase in sensitivity, astronomers have basically two options. One is to make larger telescopes. The cost of making large mirrors rises prohibitively with size, however, so alternatives have been sought. The prototype of these is the Multiple Mirror Telescope on Mt. Hopkins in southern Arizona, which has six mirrors designed to act as one and to combine their light into a single image. This telescope, which is operated by the Smithsonian Institution and the University of Arizona, has the total light-gathering power of a single 4.5 meter (176-inch) mirror at much less cost. Larger versions are in the planning stage. One of these would have four 7.5-meter (300-inch) mirrors, another would be a segmented 10-meter (400-inch) mirror consisting of 36 hexagonal pieces of glass that would be realigned 100 times a second.

The advantage of the larger telescopes will simply be that they will collect more light. A 10-meter telescope will collect four times more light than the 5-meter (200-inch) telescope on Mt. Palomar. This means that an observation requiring a certain number of photons can be made four times more quickly, or that fainter objects can be observed. The actual images made by a larger telescope will not, however, be appreciably sharper. Images made by optical telescopes on earth are blurred because of the erratic motions of cells of air in the earth's atmosphere. This effect can be seen without the aid of a telescope. It causes stars to twinkle. The twinkling of stars inspires poetic thoughts in some of us, but not in astronomers who want sharp images. Because of twinkling, the 5-meter telescope on Mt. Palomar usually produces images no sharper than a 10-centimeter

(4-inch) telescope. Astronomers have devised ingenious techniques to reconstruct a sharp image from a blurred one, but they are not practical for faint objects.

The obvious solution is to put the telescope in a place where the stars don't twinkle, that is, in space, above the turbulent, troublesome atmosphere. A 2.4-meter (94-inch) telescope in space could make images ten times sharper than the best telescopes on earth, and it could see objects fifty times dimmer. The Romanian rocket pioneer Herman Oberth pointed out some of the advantages of such a telescope in 1923, but at that time it could only be considered science fiction, and the idea was largely ignored. When Lyman Spitzer, Jr., made a similar suggestion twenty-three years later, V-2 rockets were a reality, and the possibility of putting a telescope in space could not be ignored. But it was, by almost all astronomers. Spitzer, however, is not a man to treat a good idea lightly.

One evening in 1930 Frederick Boyce, a science teacher at Phillips Academy in Andover, took a group of students into Boston to hear a lecture by Henry Norris Russell, who was then the director of the Princeton University Observatory. Over fifty years later, one of those students remembered the lecture not so much for its content as for the effect it had on him. "I don't remember [the lecture] very clearly," Spitzer said, "but I do remember that I was impressed by it at the time." That lecture, together with the influence of Boyce, an excellent teacher, turned Spitzer toward a career in astronomy.

Six years later, after graduating from Yale and spending a year at Cambridge University in England, he would be doing graduate work at Princeton under Russell. In 1938 he received a Ph.D in theoretical astrophysics. It was the first such degree ever awarded by Princeton. After a year at Harvard as a postdoctoral fellow under Harlow Shapley, Spitzer returned to Yale as an instructor in physics. During World War II, he served with the National Defense Research Committee, for whom he worked on undersea warfare research. "We sank submarines on the 64th floor of the Empire State Building," he said. After the war, he returned to Yale briefly before accepting the directorship of the Princeton Observatory, the position Henry Norris Russell had held when Spitzer had first heard him speak seventeeen years earlier.

Before the end of World War II, the United States Air Force, with an eye to the effectiveness of the German V-2 rockets, asked Cal

Tech aerodynamicist Theodore von Karman to investigate the importance of missiles in the future, from both a military and technological point of view. Von Karman suggested, among other things, that satellites in orbit around the earth were practical and had enormous military and scientific potential. The Air Force commissioned another study, called Project RAND (for "research and development"). As part of this project, Spitzer was asked to do a study of the scientific uses of a satellite.

"I wrote a brief paper on the scientific uses of satellites of different sizes," he recalled. In it he listed three general areas where an extraterrestrial observatory could be effectively used. The first two involved the use of small telescopes to study ultraviolet radiation from the sun and stars. Ultraviolet radiation is, as the name suggests, electromagnetic radiation having frequencies larger than violet light, which is at the upper end of the visible portion of the spectrum. Ultraviolet radiation at frequencies just higher than that of violet light is beneficial if taken in the proper doses; it builds up vitamin D. In larger doses it can cause sunburn or even skin cancer. Higher-frequency ultraviolet radiation, which merges gradually into the low-frequency part of the x-ray band, carries more energy per photon and can be very harmful to living cells. Life on earth as we know it would not be possible if the atmosphere did not absorb the far ultraviolet radiation from the sun. Atmospheric absorbtion of ultraviolet radiation requires, however, that astronomers place their telescopes on rockets, balloons, and satellites if they wish to observe ultraviolet radiation from the sun and stars; hence Spitzer's recommendation. The telescopes and detectors used for ultraviolet astronomy are essentially the same as those used for optical astronomy, except that they have to be made of lightweight materials since they must be lifted into orbit.

Astronomers need to make ultraviolet observations for two reasons. First, objects that have temperatures between about 10,000° Celsius and a million degrees give off most of their energy in the ultraviolet. Massive stars especially cannot be adequately understood without observing them in ultraviolet light. And even in lighter stars such as the sun, the hot outer layers can be studied only through ultraviolet and x-ray observations. Another fundamental reason for extending our vision into the ultraviolet range is that many elements and molecules in stars and interstellar space strongly emit or absorb radiation at ultraviolet frequencies.

In his report, Spitzer also discussed putting a large reflecting telescope on a satellite to be used for both optical and ultraviolet observations, and he outlined some of the scientific programs that could be attacked with such a telescope. It would be almost twenty years before even a small ultraviolet telescope would be flown; and by the time a large optical and ultraviolet space telescope is launched, it will have been at least forty years. Incredibly, Spitzer has been a persistent champion of the idea throughout those four decades.

The launch of NASA's Edwin P. Hubble Space Telescope is scheduled for August of 1986. It has been named in honor of the astronomer who led us beyond the galaxy into the expanding universe with his pioneering work on the first large telescope to be constructed in this century, the 2.5-meter (100-inch) telescope on Mt. Wilson. It is called simply "Space Telescope," without reference to the wavelengths of radiation the instrument will detect, from the middle of the ultraviolet through the optical to the near infrared. This reflects the traditional thinking among many astronomers that "telescope" means "optical telescope," whereas any other type of telescope or observatory has to be qualified with a prefix, as in the *Einstein* x-ray observatory or the Gamma Ray Observatory or the Infrared Astronomical Satellite.

The Space Telescope will orbit the earth at an altitude of 500 kilometers (300 miles), well above the the obscuring layers of the atmosphere. Its 2.4-meter (94-inch) mirror and complement of instruments will allow the Space Telescope to make exquisite, high-resolution images of star systems and clouds of gas, to determine the positions of sources precisely, to detect extremely faint objects, and to make detailed spectral measurements.

The astronomical community is understandly looking forward to launch with great enthusiasm and anticipation. It was not always so. For at least a decade after Spitzer made his recommendations in the Project RAND report, he talked the idea of a space telescope up among his colleagues, but few of them listened. For one thing, the 5-meter reflecting telescope on Mt. Palomar, which was dedicated in 1948, was so superior to any previous telescope that most astronomers could not justify the expense of putting a large, reflecting telescope in space, if indeed such a venture was technically possible. After all, the very idea of satellites was, at that time, still on the drawing boards.

Meanwhile, a group of scientists at the Naval Research Laboratory was pushing into the new frontier. In October of 1946, Richard Tousey and colleagues used a captured German V-2 rocket to obtain the first ultraviolet observations of the sun. Less than three years later, a Naval Research Laboratory team led by Herbert Friedman used a V-2 to make the first detection of x-radiation from the sun. In 1957 the same group, having switched to the more reliable Aerobee rockets developed by another space-science pioneer, James Van Allen of the University of Iowa, detected ultraviolet radiation from a star other than the sun for the first time.

About that time, Spitzer arranged to have Van Allen pay an extended visit to Princeton, partly in hopes that his presence would help to stimulate enthusiasm for space research among the Princeton faculty. Spitzer was not disappointed. One of his colleagues, Martin Schwarzschild, was interested in measuring the turbulent motions on the surface of the sun. Viewed with a powerful optical telescope, the sun does not appear smooth; rather, it has more the character of a pan of boiling oatmeal. This turbulence is caused by rising and falling columns of gas beneath the surface. The nature of these motions is not well understood; Schwarzschild wanted to study them in detail but was frustrated by the fuzziness that the atmosphere introduces in solar images made by the best ground-based telescopes.

"At lunch one day," Spitzer recalled, "I described Schwarzschild's interest in the sun, and Van Allen said, 'Why don't you send a telescope up in a balloon?' . . . We talked about it and it seemed practical. I was interested in promoting a program here at Princeton that would lead in time to a large space telescope. This balloon telescope seemed to be an important step in that direction."

Spitzer helped Schwarzschild to raise money for the project from the United States Navy and the National Science Foundation. In a series of balloon flights in 1957 and 1959, Schwarzschild's group sent an automated 25-centimeter (10-inch) optical telescope called Stratoscope I aloft to a altitude of 24,000 meters. Stratoscope I's exceptionally clear photographs of the sun revealed details of the solar structure that could be seen by the best ground-based telescopes only on rare occasions when atmospheric turbulence was at a minimum. In the following decade a 90-centimeter (36-inch) telescope called Stratoscope II made photographs with a resolution of a tenth of a second of arc, or ten times better than is normally possible with a ground-based telescope. The advantages of placing a telescope above the atmosphere had been clearly demonstrated. Incidentally, Stratoscope II had also carried Low's first infrared bolometer system aloft to make infrared observations of Mars.

By now it was the mid 1960s and Spitzer was beginning to make headway with his concept of a large space telescope. Shortly after the establishment of NASA in 1958, a program of four Orbiting Astronomical Observatories was planned. These observatories would use four identical spacecraft outfitted with different instruments for measuring ultraviolet radiation from stars or from gas between the stars. A Princeton group led by Spitzer proposed a 36-inch telescope with instrumentation that would allow it to measure ultraviolet spectral lines. This experiment, which was the most complex, was selected for the fourth mission.

The first mission failed because of electronic malfunctions after only a few days in orbit. The third mission was also a disaster: the rocket failed and the spacecraft was dumped into the Indian Ocean. The second mission, launched in 1968, and the fourth mission, launched in 1972, were both succcessful. They provided data which has radically changed our understanding of conditions in the interstellar gas. It is not spread out uniformly, as scientists had supposed; rather, there are dense gaseous clouds of molecules of various types, permeated with vast tunnels of hot, low-density gas formed by exploding stars. In addition to their scientific value, these missions, along with other satellite missions, helped to establish the technological base necessary for making a large space telescope feasible.

In the meantime, Spitzer had been laboring to establish the grassroots support that would be necessary if a space telescope were ever to get into orbit. It was, as Spitzer recalled, "a very gradual process." As early as 1962 he had met with a senior group of scientists convened by the National Academy of Sciences at Iowa City, Iowa, to lay out a program for space science. Spitzer was there to promote the concept of a large space telescope, but he was not encouraged. "When we first talked about it at the Iowa summer study, it was clear that it was not going to get such a tremendous endorsement from that group." He was right. The group identified a space telescope as a natural "long-range" goal of the space astronomy program— in effect, exiling it to the back of the line with faint praise.

Things went better three years later at a similar study at Woods Hole, Massachusetts. By then, the *Apollo* program to land a man on the moon was in full swing and all of space science was benefitting. A space telescope, along with the program that would lead to the High-Energy Astrophysical Observatories, were given strong recommendations. After that, Spitzer said, "I was sufficiently enthusiastic and hopeful to go on the assumption that in due course it was going to get somewhere."

Following the recommendation of the Woods Hole group, the National Academy of Sciences appointed Spitzer as chairman of a committee to establish the scientific goals and usefulness of a large space telescope. A summary of the committee's report, which outlined in detail what could be done with a large space telescope and its advantages over ground-based instruments, was published in *Science* magazine in 1968; the full report was issued in 1969. It represented an important step in selling the concept of a space telescope. "More important than that," Spitzer said, "was a series of meetings we had with astronomers all over the country from time to time on different topics. The net effect of those meetings was to spread the gospel, to expose other people as to what the astronomical possibilities were and gradually gain converts and support from the astronomical community. So, by the time we completed our study, I think there was reasonable support from astronomers."

The support was little more than reasonable. Certainly, the endorsement contained in the influential Greenstein Committee Report cannot be described as fervent. Surveying the needs of astronomy for the 1970s, the committee supported the general concept of a space telescope but still considered it to be an idea for the fairly distant future. It was ninth on a list of nine recommendations, of which only the top four were considered top-priority items.

NASA responded somewhat more encouragingly, possibly because the *Apollo* program was winding down and the agency was seeking scientific uses for the proposed Space Shuttle. In the fall of 1971, feasibility studies for the Space Telescope were begun at NASA's Goddard and Marshall Space Flight Centers. A small group of scientists headed by Nancy Roman, who was then head of the Astronomy and Relativity office at NASA headquarters, provided scientific guidance. In 1972 NASA designated Marshall as the lead center for the project. Charles Robert O'Dell, who had been named director of the University of Chicago's Yerkes Observatory at age 29, accepted the position of project scientist.

O'Dell was born in 1937 in the 'Little Egypt' section of southern Illinois, where he grew up. He could trace his interest in astronomy back to the sixth grade, when his teacher asked the class to write a one-page paper on what they would want to be doing twenty-five years in the future. "I can remember writing that I wanted to be an astronomer using the 200-inch telescope. Now that I look back on it . . . that was just after the 200-inch became operational, so they must have been getting lots of publicity at the time." Having mapped out his future, O'Dell set about to make it happen. He graduated from Illinois State University with a bachelor of science in education and three years later, in 1962, received a Ph.D in astronomy from the University of Wisconsin. He spent a year as a Carnegie fellow at Cal Tech, only to discover that the 200-inch telescope was off limits to Carnegie fellows. After a year at Cal Tech, he spent a year and a half as an assistant professor at the University of California, Berkeley, before going to the University of Chicago, where in 1966 he became the director of the Yerkes Observatory and a year later, chairman of the Astronomy Department. In this capacity he returned to Cal Tech as a distinguished visitor and was allowed to use the 200-inch telescope, eight years ahead of the schedule he had set for himself in the sixth grade.

O'Dell was an early convert to the idea of the Space Telescope, so when he was given the opportunity to help make it happen, he took it, even though it meant giving up the security of academic life. "He was a full professor, active scientist, very distinguished young man," John Bahcall of the Institute for Advanced Study remarked, "and he realized that this program needed somebody. Lyman [Spitzer] wasn't willing to take over the activity of Phase B [the preliminary design phase]. Bob did. Bob made the sacrifice . . . Bob eventually gave up his tenured positions and dedicated himself to this full time, at great personal sacrifice, not knowing whether or not it would happen but feeling he had to try."

As project scientist for the Space Telescope, O'Dell had to have one foot planted in each of two worlds, the world of the astronomers and the world of NASA. In the design phase, he had to coordinate the dreams of the scientists with the realities of the engineers, and he had to convince both the astronomers and NASA that the Space Telescope was the best way to realize those dreams.

"The first obstacles were the astronomers themselves," he told us.

Many astronomers felt that the money for the Space Telescope, originally projected to be around $300 million, would be better spent for a dozen or so ground-based telescopes than for one space telescope. "Initially there was a great resistance among my colleagues, the traditional optical astronomers, to a program that spent all that much money on one instrument." O'Dell feels that they were just defending the status quo, that they were unwilling to admit that times were changing, even for optical astronomy. "The optical astronomers are, after all, a pretty conservative group. We can see that today, when they still want to build 7-meter and 10-meter ground-based telescopes and think they'd be just wonderful, whereas they will hardly improve on our ability to make discoveries."

"This mental inertia of the ground-based optical astronomers was the first thing to overcome," he said. "That's never really been done," he admitted, "because I'm sure that people would only support it [the Space Telescope] because they know that if you continue to refuse to support NASA programs, it would not result in more money being spent for other parts of astronomy. Not spending $100 million in NASA on a Space Telescope doesn't mean that $100 million is going to appear in the National Science Foundation's programs. So I think it was finally a combined recognition of this different color of money, together with a genuine recognition that the Space Telescope would do things that *no*," he emphasized the 'no,' "ground-based telescope could do, that finally won people over."

The next hurdle was to sell the program within the agency. A successful new program within NASA must make it over several hurdles. Many ideas for new programs are submitted to the agency, from individuals or groups on both the inside and outside. A few of these are subjected to feasibility studies, that is, studies which give some idea as to whether the proposed program would be practical or not. This is called Phase A. A small fraction of the Phase A studies advance to Phase B, the preliminary-design phase, during which a basic design, together with a schedule and approximate costs, are developed. The feasibility and preliminary-design phases are usually paid for from general funds set aside for such studies and, to some extent, by aerospace contractors who hope that their design will be selected if the program gets funded. Only a few projects make it through to the next phase, design and development, where they are included by name as a new project in NASA's budget. New projects must be accepted by the Office of Management and Budget and by Congress. Still more projects fall by the wayside at this point.

"We were in competition with lots and lots of other good ideas," O'Dell recalled. "It's easy to beat out the not-so-good ideas, but there were a lot of other programs that wanted to get started, too. Of course the planetary people in those days were really riding high. That was just the start of the period when one was able to do significant planetary obervations and even exploration," he explained, "so the funding was really starting to build up where it had been fairly low before. We were in direct conflict there." The planetary people won this particular battle; the Space Telescope program had to wait in line until after *Viking* had landed on Mars, in 1976.

But as early as 1974, O'Dell and the other boosters of the Space Telescope within NASA had marshaled enough support to get the project in line to start design and development on July 1, 1975. For a brief period, prospects were bright for a launch sometime in 1982, but storm clouds were rapidly building up. The economy was suffering from the oil crisis, and NASA was suffering from a lack of support in Congress and the White House, largely because of cost overruns in the *Viking* program. The High Energy Astrophysics program was cut back drastically during this period, and NASA officials were uncertain as to how to present a program as expensive as the Space Telescope to Congress. They decided to include it as a specific item in the budget for the fiscal year 1975 (which would begin on July 1, 1974), even though the project was only in the preliminary design phase and could have been covered by general funds. This proved to be a serious tactical error.

The House Appropriations Subcommittee responsible for NASA appropriations deleted all funds for the Space Telescope. Because the subcommittee was also responsible for the National Science Foundation budget, its members were aware that funding had recently been approved for the Very Large Array of radio telescopes. They were also aware of the Greenstein Committee's ranking of projects. In its report, the full committee on Appropriations used this ranking as part of its argument for eliminating funds for the Space Telescope: "The Committee notes that the LST [Large Space Telescope] is not among the top four priority telescope projects selected by the National Academy of Sciences, and suggests that a less expensive and ambitious project be considered as a possible alternative." "If it had been left that way," O'Dell said, "the program would have been dead before it got started." O'Dell, as a civil servant, was limited in what he could do to save the project. Eloquent and energetic advocates from the astronomical community were desperately needed. Fortunately for the Space Telescope, Spitzer and John Bahcall stepped forward to fill these roles.

Spitzer was by then a senior statesman of astronomy, experienced in dealing with congressional committees; moreover, he had been promoting the Space Telescope for twenty-seven years and was not ready to give up the fight. Bahcall, of the Institute for Advanced Study in Princeton, had been associated with the project since 1972 in an advisory capacity. Since obtaining his Ph.D from Harvard in 1960, he had risen quickly into the top ranks of astronomy and established a formidable reputation as an excellent scientist, a tough debater, and a dynamo whose prolific scientific output seemed unaffected by the almost continual round of speaking engagements and committee meetings that sap the energy and creativity of many successful scientists. When we telephoned him to arrange an interview, he warned that "I'll be a moving target," since his schedule was full of talks and meetings at Cal Tech and UCLA. Finally, we met with him one morning before breakfast (before our breakfast—he had eaten much earlier) in a Pasadena motel.

Bahcall grew up in Shreveport, Louisiana. His father was a salesman and his mother has worked as a social worker and in sales. "Both were intelligent people who liked to read and think," he said. After finishing high school in Shreveport, where he excelled as a debater, Bahcall entered the University of California at Berkeley. "I entered college thinking I would be a rabbi," he recalled. "I never believed I was going to be a scientist until my junior year at Berkeley. I was a philosophy major, and I had not had any science in high school.

Bahcall might never have taken a science course in college, either, if it were not for a graduation requirement. "I took the physics course for physicists," he said, chuckling at the memory. "That was very hard, very challenging, but very interesting and that just sold me right there." He crammed all the physics and mathematics requirements into the next two years and graduated with a bachelor's degree in physics. After Berkeley, Bahcall went to the University of Chicago, obtained a master's degree in a year, and then went to Harvard, where he received a Ph.D. in 1961. After a year of postdoctoral research at Indiana University, he moved to Cal Tech, at the invitation of William Fowler. While there, he began making regular visits to the Weizmann Institute of Science in Israel. During one of these visits he met Neta Assaf, a graduate student who was just finishing her master's degree. "The first ten times I asked her out she said she was too busy, that she was preparing an experiment," he laughed. In 1966 they were married. Neta Bahcall now has a Ph.D in astronomy and is in charge of the program for allocating time on the Space Telescope to guest observers.

John Bahcall became interested in the Space Telescope in the early 1970s when the Greenstein Committee's report was being written. After NASA hired O'Dell and began the preliminary design studies, Bahcall remembered, "I wrote a letter to Bob O'Dell and Nancy Roman saying . . . I was very enthusiastic about the Space Telescope and I would be happy to help in any way I could . . . And Bob wrote me back a letter, sometime later, saying 'You know, we need theorists in the working group, and why don't you apply for that?' "

Bahcall did apply and was selected, along with Spitzer, to join a twelve-member working group chaired by O'Dell. "For the first year or two I just worked on technical problems," he explained. "But when the program got into trouble in '74, I was close and committed enough, together with Lyman, that we felt we had to quickly do a lot of lobbying." During the next few years Bahcall spent, by his own account, more than half of his time lobbying for the Space Telescope.

The first lobbying priority was with the astronomers themselves, especially those on the powerful Greenstein Committee. With the help of O'Dell and Spitzer, who were members, Bahcall persuaded the committee, including the influential Greenstein, to abandon their earlier weak position and to present a united front in support of the Space Telescope as a top-priority project. Although this support was probably never broadly based, it was enough to patch up one of the major holes in their argument, namely, that the astronomers themselves were not particularly excited about the Space Telescope. This done, Bahcall and Spitzer turned their attention to Congress. They shuttled back and forth between Princeton and Washington during July and August, "explaining," in Bahcall's words, "to senators, to congressmen, to congressional staffers, and to anyone else who would listen the importance of the program and the necessity of maintaining momentum in the project. We were joined by a number of colleagues at greater distances from Washington who telephoned, wrote, and sometimes visited representatives from their home areas." It worked. The Senate Appropriations Subcommittee voted to restore funds for planning studies in fiscal 1975; these funds were cut in half, to $3 million in a House–Senate compromise.

The supporters of the Space Telescope rejoiced at this narrow escape, and the working group pushed forward. Their hope was to finish the preliminary design by mid-1976 and begin development with the substantial funding that would be available in fiscal year 1977. This hope was dashed with the appearance of NASA's draft of the budget for fiscal year 1977. The Space Telescope did not appear. Apparently NASA intended to continue funding it at a low level out of general funds as a project that was still in the preliminary design phase. What had happened?

"Well, a lot of things happened," Bahcall explained. "First of all, we believed that our problems in getting the Space Telescope funded had to do with a lack of commitment to basic science at the highest levels of NASA. NASA's mission is not science, as you know. NASA's mission, as perceived by NASA administrators, . . . is primarily to keep NASA employees employed, and not necessarily to support basic science. It doesn't exist in the structure. So," he continued, "they were not really committed to the program and they . . . gave rather lukewarm support to it. I, because of good contacts with some NASA people and other people on the outside, knew what was happening. I was informed in a way that was not," he paused to choose his words carefully, "direct. I knew that . . . the scientists had to make a large splash or NASA would trade us off for something that was going to help the aerospace industry and the NASA centers and NASA's other constituencies more than it would help the scientists.

"It was important," he emphasized, "that some people in NASA were very committed to science and were interested in helping us. We had to have information about what was happening or we couldn't be effective. We had to have prior information."

Since the budget was out of NASA's hands and into the hands of Congress, any change would have to be made on the Hill. This is an unusual but not unheard of step. Bahcall, Spitzer, and their colleagues began calling, writing, and visiting key congressional committee members again.

Their contacts among lobbyists for the aerospace industry proved valuable. "We worked very closely with people in the aerospace industry, who were very good lobbyists and knew really what each senator was doing on the weekend and also knew what the top NASA administrators were doing," Bahcall said with amusement. "Who saw who, when, and what were the arguments that were appropriate. Lyman and I could get appointments with these people."

He explained how he would go about setting up a meeting with a senator or other highly placed person. In some cases he could simply call and ask for an appointment. More frequently, though, he used political contacts. For example, Carl Kaysen, then the director of the Institute for Advanced Studies, put him in touch with an old friend and early ally of the Space Telescope, Senator Charles Mathias of Maryland. Barbara Sigmund, mayor of Princeton, arranged an appointment with her mother, Representative Lindy Boggs of Louisiana. An old high school friend from Louisiana arranged an appointment with a Louisiana senator.

"At that time," he explained, "scientists in Washington were not very common, so people were willing to spend fifteen minutes or half an hour with a lone scientist where they might not be willing to spend that time with an aerospace lobbyist who comes in week after week with different problems . . . So the aerospace people used us and we used them. They knew that Lyman and I could get to see people that they couldn't, and so they supplied us with data—how much money would likely be spent in each state and arguments that were relevant to this person's constituency."

"In all cases we would try to present the excitement and uniqueness of the scientific program. Because that's really why we were interested in it, too. But there were congressmen who were just not interested in that at all, who had made up their minds. Crucial congressmen made up their minds to support the program once they had heard about it from their aides, just because there was a favorable economic impact on their district. Maybe they heard one sentence about the program and it was fine and they bought it—it employed five hundred people in their district. There were other congressmen who did it because they were doing a favor for a second congressman interested in it. There were people who were interested in it because of the technology spin-off or the belief that a strong space program helped our national defense effort."

One persistent question raised by critics of the space program was whether the money wouldn't be better spent curing some of the problems here on earth. "There were people who asked us," Bahcall said, " 'Won't the stars be there a decade from now? Why do you need this now? Isn't it more important now to eradicate the rats in the inner cities?' That, I think, was for me the most difficult question. Because if it were a simple trade-off of eradicating the rats in the inner city or funding an astronomical experiment, eradicate the rats in the inner city, now. No one questioned that. But it wasn't that kind of trade-off. Because the monies were in different budgets. The question really was a trade-off between one NASA-supported program or one technical program versus another . . . There would be a minimum amount of money spent in space anyway, and we were arguing that it was a wise investment in the long run for the government, the nation's economy, and the people in it, for their intellectual enrichment and for their economic benefit via technological spin-offs, to support the Space Telescope."

Bahcall went virtually everywhere in search of congressional support, including the women's lounge near the floor of the House of Representatives. There he interviewed Representative Lindy Boggs, who then helped him line up appointments with other members of Congress. He had originally planned to see her in her office, but "it turned out that she was on the floor of the House and there was some vote which she had to stay nearby for . . . About the only quiet place around there was this lounge which is associated with the women's restroom. I guess it's not unusual to take guests in there," he laughed, "but I found it a bit intimidating."

Bahcall found that the home-state touch carried considerable weight with politicians. "It was very important that people from that person's home state wrote, called, or contacted them in some way," he said. "I think we had, in each of the states that were relevant, astronomers to write, call, and in some cases visit with us. Whenever possible, we went with somebody from the home state. That made a lot of difference."

Officially, NASA had to watch all this activity from the sidelines, since the administration had approved the budget without a provision for developmental funds for the Space Telescope. Bahcall was never certain whether the NASA administrators were cheering for or against the efforts of the scientists to save the Space Telescope; he had an uneasy feeling that they were not on his side. "My friends [unidentified supporters of the Space Telescope within NASA] felt that the program had to be saved by the scientists, that it wouldn't be saved by NASA . . . There was another, I would say a more team-player person at NASA, who, at the time the program was in most trouble, urged Lyman and myself not to lobby, not to stir up the scientists but to leave it to NASA, because NASA's good will was necessary for the support of science—that apart from NASA's Space Telescope, there were other programs that needed NASA's good will . . . In the end Lyman and I decided to ignore that advice. It conflicted with other advice and I think it was the right thing to do, because once NASA saw that there was widespread scientific and professional support for the Space Telescope, they weren't mad at the scientists . . . They were grateful for stirring up support for any NASA program and they hopped on the bandwagon. Now NASA claims it is their number-one scientific priority. It certainly wasn't then."

In the end, NASA's reluctance and the opposition of a few key representatives and senators, who had consistently opposed the Space Telescope, defeated the scientists' attempts to get funds for the development of the Space Telescope added to the 1977 budget. But though they had lost the battle, they won the war. NASA was authorized to begin the selection process for the scientists and contractors who would work on the telescope, so that development could begin as soon as possible in fiscal year 1978.

"That was thrilling," Bahcall recalled. "That was a thrill which was shared with people like Lyman, Margaret Burbidge, George Field, George Wallerstein, lots of people around the country joined together in that. It was a real thrill to have the scientists behind the program. It guaranteed good science."

20

The final approval of the Space Telescope program by Congress was understandably a high point for Bob O'Dell. "I have to admit that the most fun," he said as he reflected on the eleven years he spent as project scientist for the Space Telescope, "was pushing it through all the resistance within the agency, in the astronomy community, and finally on the Hill, and experiencing that feeling of creation that came with seeing the project getting started."

"From then on," he continued, "it has been satisfying to see it come into existence, to carry out what one is committed to do."

His commitment was extraordinary, according to John Bahcall. In his capacity as project scientist, O'Dell was chairman of the Science Working Group—eighteen scientists who were responsible for the scientific requirements of the Space Telescope. "Bob worked with a bunch of prima donnas," said Bahcall, who was himself a member of the group. "There were a bunch of us on that committee and we all had different view points. I think Bob did a magnificent job. There wouldn't have been a space telescope without him . . . I think his unselfishness in forwarding the goal, which he believed to be in the interest of science, was just extraordinary . . . He's very technically able, and above all else he, by his dedication, his conviction, he's an inspiration to everybody in the program. I wouldn't have been there without him."

But in spite of the dedication of O'Dell and many others like him, the Space Telescope program encountered further, serious problems. In August of 1979, the launch date was set for early 1983 and the cost was estimated at $440 million. A little over a year later, the launch had slipped two years, to early 1985, and the cost had risen by 15 percent. By 1985, the launch had slipped another year and a half and the cost had more than doubled to over a billion dollars. The high price tag of the Space Telescope has dismayed astronomers and politicians alike. Astronomers worry that it is sucking up funds for other important astronomical projects.

The reasons for the escalating cost, James Beggs, the NASA administrator, told a congressional investigating committee, were that the Space Telescope "is the toughest job NASA has ever tried to do," a job whose technical complexities the agency and its contractors "seriously underestimated."

The Space Telescope has three major components: the optical telescope, the system support module (that is, the spacecraft in which the telescope system is housed), and the scientific instruments, which detect the light focused by the telescope. The basic technology for each of these components was developed in earlier space science experiments, such as the Orbiting Astronomical Observatories, the *Stratoscope* project, the Orbiting Solar Observatories, and *Skylab*, and in military reconnaissance systems.

Because the Space Telescope was designed to be carried to and from orbit by the Space Shuttle, this limited the mirror size to 2.4 meters (94 inches). The mirror and alignment system had to be sturdy enough to withstand strong vibrations during launch and gravitational forces four times those on earth during landing, yet light enough for the Space Shuttle to lift into orbit. Since the Space Telescope will produce much sharper images than ground-based telescopes, much greater pointing and stabilizing accuracy is required. And it must be achieved while floating in space without anything to hold on to. For peak performance, this fine guidance system must be able to track an object with an accuracy of seven thousandths of an arc second. As O'Dell remarked, "It's a horribly, wonderfully complex system." He likened it to asking someone to drive a car by remote control from Los Angeles to Boston and park the car within six inches of an arbitrary spot. The system uses gyroscopes, reaction wheels, star trackers, and sensors which lock onto a pair of guide stars in the vicinity of a target and update the position of the telescope every second.

"Early on, it looked like just building the optics was going to be the problem," O'Dell told us, "but then, through a series of test pieces, basically a demonstration program, that question was resolved." The ITEK Corporation asked Corning Corporation to make a mirror blank in which the front and back plates are fused to a core that is honeycombed like an egg crate. ITEK used this blank to make a light, high-quality test mirror, laying to rest one of the worries of the designers. The General Dynamics Corporation constructed a graphite epoxy framework that was light and strong enough to hold the optical system in alignment during launch and landing, one that would not expand or contract too much in response to the temperature extremes it might encounter. This model established that the new material could be used in the design of the Space Telescope. The Martin-Marietta Company tested the reaction wheels proposed to control the pointing of the Space Telescope on a specially designed table, one having the characteristics of a telescope floating in space, and demonstrated that they could control the telescope with the required precision.

By the time Congress approved the development of the Space Telescope, these successful tests had been made; thus, O'Dell and the others could enter the hardware or building phase with confidence. We asked O'Dell if he ever lost that confidence, if ever there was a time when he thought that the technical problems were going to be too great to overcome.

"No," he replied. "ST has always to me been something that was feasible. I never saw a technical problem I didn't think that we could resolve . . . There were no inventions on the ST program. It was just doing things bigger and better."

Marshall Space Flight Center was given overall responsibility for the program, as well as specific responsibility for the optical telescope assembly and the system support module, but not for the scientific instruments and the postlaunch operations. Those went to Goddard Space Flight Center. This arrangement was characterized by Beggs as "an unusual management structure" and was blamed for much of the trouble that followed. This was about the same time that NASA set up a similar management structure for the Infrared Astronomy Satellite, with the Jet Propulsion Laboratory and Moffat Field centers sharing responsibility. Given Goddard's experience in the development and operation of space telescopes aboard the Orbiting Astronomical Satellites and the Explorer satellites, and their recent loss of the Infrared Astronomical Satellite project to Moffat Field, it is not surprising that NASA administrators gave a piece of the Space Telescope pie to Goddard Space Flight Center. This appears to be a recurring pattern: NASA administrators try to keep as many groups as possible happy by diffusing responsibilities, with little regard for future managerial problems. The result, inevitably is poor communication.

This problem was exacerbated by the rapid turnover of directors at the Office of Space Sciences at NASA headquarters (five directors in six years), as well as a division of responsibility among prime contractors. Lockheed Missiles and Space Corporation had responsibility for the systems support module, whereas Perkin-Elmer Corporation was put in charge of the optical telescope assembly. Yet another group, the European Space Agency, was to provide the solar panels that would power the Space Telescope, as well as one of the scientific instruments. The European connection was established through the mutual interest of NASA and the European Space Agency with the enthusiastic support of Congress; it had, in O'Dell's words, been "a very positive part of the Space Telescope program."

Toward the end of 1980, it was clear that Perkin-Elmer was going to produce a very high-quality mirror. It was also clear that they were not going to do so on time. In addition, work on the fine guidance sensors was not going well. An extensive review also uncovered problems with the development of both the scientific instruments and the support system module. The cost estimate was increased by 15 percent and the launch date was set back a little over a year, to early 1985.

On December 5, 1981, at Perkin-Elmer, the mirror, which had been polished twenty-six times until it was the most precise large mirror ever made, was put into a stainless steel vacuum chamber manufactured by Mill Lane Engineering. The air was sucked out by huge pumps, and then small slugs of aluminum were released into the chamber. These were instantly vaporized by electron-beam guns. A turntable rotated the mirror slowly to ensure that the aluminum molecules would settle evenly on the glass. Within two minutes, a reflective layer of aluminum a few millionths of a centimeter thick had coated the glass. The same process was used to apply a thin overcoat of magnesium fluoride to guard against oxidation of the aluminum surface and to enhance the reflectivity of the mirror to ultraviolet radiation. Tests of the mirror indicated that it exceeded specifications and was the finest mirror ever made. If the Pacific Ocean were as smooth as the Space Telescope mirror, the largest wave would be only a thousandth of an inch high.

While the technical excellence of Perkin-Elmer was beyond doubt, serious questions were soon to be raised about management practices within Perkin-Elmer and NASA. The project was slipping further behind schedule every month, and neither Perkin-Elmer nor Marshall Space Flight Center seemed to know what to do about it. In the meantime, dust was allowed to accumulate on the mirror, producing possibly irreversible contamination. The latches that would hold the scientific instruments in position chafed during movement; this might have prevented proper realignment after the instruments were removed for servicing. Also, it was not clear that the fine guidance sensors or the pointing control system would work up to specifications, and there were problems with the solar arrays. New managers were brought in; tiger teams were formed; Congress investigated. The launch was pushed back to late 1986 and the estimated total cost increased to $1.2 billion.

Beggs admitted to the House Subcommittee on Space Science and Applications that the management structure had failed to do its job. "Good management recognizes troubles early and takes appropriate action," Beggs said to Congress. "We underestimated the time, money, and people it would take to deal with the difficulties and uncertainties which were inherent in the program from the beginning but went unrecognized." Heads rolled at NASA headquarters, Marshall Space Flight Center, and Perkin-Elmer, and NASA's on-site supervision of Perkin-Elmer was intensified.

The dust on the mirror was removed by blowing dry nitrogen across the surface and using vacuum hoses to remove the tiny particles; then the telescope was sealed to prevent a recurrence of contamination. In the meantime, Perkin-Elmer had taken extra precautions to ensure that molecular diffusion from the graphite epoxy frame onto the mirror would not be a problem. The fine guidance sensors, which used a system of prisms monitored and controlled by a computer, proved to be one of the most difficult technical challenges, but in the end they worked better than NASA's specifications. The troublesome latches were coated with tungsten carbide and cobalt; this tough material proved resistant to the chafing problem that had plagued the softer aluminum oxide coating. In late 1984, in keeping with the revised schedule, the optical telescope assembly was completed. The twelve-ton package was shipped by a "Super Guppy" C5A transport plane to the west coast for integration with the spacecraft and further testing at Lockheed. If the project stays on schedule, the Space Telescope will be shipped through the Panama Canal to Cape Canaveral in March 1986, in time for an August 1986 launch.

The Space Telescope is expected to remain operational for at least fifteen years, and possibly much longer. The key to its long lifetime is the Space Shuttle, which will be used to place it in orbit and to service it as the need arises. With the Space Telescope, astronomers will, for the first time, be able to have their cake and eat it, too. They can put a telescope above the troublesome atmosphere, yet if something goes wrong, it can be repaired by an astronaut or, if necessary, returned to earth. As better scientific instruments become available, it will be a relatively simple matter to install them at a fraction of the cost of a new satellite. Thus, the success of the mission is virtually guaranteed. The Space Telescope project will therefore be more than the most effective optical and ultraviolet telescope ever; it will represent the beginning of a new age in astronomy—the establishment of permanent observatories in space.

During its first month in orbit, the telescope will be switched on, pointed, and focused, while the solar arrays will be checked for power delivery and the antennas will be checked to ensure that the communication links are established. For the next five months the telescope will be used for scientific observation and a long series of tests to thoroughly evaluate the performance of each of the scientific instruments.

The scientific instruments that detect the radiation concentrated by the telescope are mounted behind the primary mirror. Light entering the telescope is reflected by the primary mirror to a secondary mirror, which reflects and focuses the radiation back toward the primary mirror; this radiation passes through a hole in the primary mirror and comes to a focus behind it. Small mirrors behind the primary mirror deflect the light so that part of it falls on each of the fine guidance sensors and part on the detectors. The fine guidance sensors, in addition to guiding the telescope, will be used to provide accurate positions of stars.

The other five scientific instruments can be used for a variety of observations. The *wide-field/planetary camera* will be the workhorse of the Space Telescope because of its versatility. It uses an array of charged coupled device detectors that are sensitive to a wide range of wavelengths, from the far ultraviolet to the near infrared. It can be operated as a wide-field camera to examine large areas of the sky, or it can be used to make high-resolution images of planets and galaxies and to search for planets around other stars. The principal investigator for the wide-field/planetary camera is James Westphal of Cal Tech.

The *faint-object camera* will be supplied by the European Space

Agency, with F. Ducio Machetto as project scientist. This instrument is designed to use the full resolving power of the Space Telescope to detect objects fainter than have ever been observed before and to make the sharpest possible images of clusters of stars, galaxies, and clouds of gas.

The *faint-object spectrograph* will be used to spread out the light from a distant galaxy into its various colors or wavelengths. The study of the spectra of stars, gas, and galaxies is one of the best ways to learn about the chemical composition, temperature, and state of motion in these objects; thus, this instrument will be one of the most popular ones on the Space Telescope. For example, it will be used to study the spectra of distant quasars and to determine the amount of helium present in galaxies when they were just forming 10 or 15 billion years ago. Since the amount of helium produced in the Big Bang depends on the density of matter in the universe, and the density of matter in the universe determines whether the universe will expand forever or collapse, the faint-object spectrograph could provide us with information concerning the ultimate fate of the universe. Richard Harms of the University of California, San Diego, was principal investigator for this instrument until he left the program and was succeeded by E. Margaret Burbidge, also of the University of California, San Diego.

For more intense sources, finer spectral resolution can be obtained. The *high-resolution spectrograph* is designed to study such things as the composition of comets in our solar system and the elements expelled into interstellar space in supernova explosions. It can also test for the existence of black holes in the center of galaxies. John Brandt of Goddard Space Flight Center is principal investigator for the high-resolution spectrograph.

The *high-speed photometer* is designed to make highly accurate measurements of the intensity of light from a star. This instrument will be used, for example, to study the rapid time variation of light from flaring stars and possibly from double-star systems containing black holes. The principal investigator for the high-speed photometer is Robert Bless of the University of Wisconsin.

The data gathered by the telescope will be transmitted up to one of the two satellites in NASA's new Tracking and Data Relay Satellite System. These satellites are in geosynchronous orbits 36,000 km (22,500 miles) above the earth. Because the only ground station set up to receive data from these satellites is in White Sands, New Mexico, the data will be beamed down to White Sands, then up again 36,000 km to a commercial communications satellite, then down again to Goddard Space Flight Center, from whence it will be transmitted 40 kilometers via a land line to the Space Telescope Science Institute on the campus of The Johns Hopkins University in Baltimore, Maryland. The first 140,000 or so kilometers of the trip that the data will take are relatively routine and uncontroversial. The last 40 kilometers have been, and still are, contentious.

21

It has been estimated that over 10,000 people have had a hand in building the Space Telescope. The complexity and power of this first permanent observatory in space is such that it will take several hundred scientists, engineers, and technicians working full time to operate it and to handle the flood of information that it will generate. The Space Telescope has the potential for observing 24 hours a day, but the low-altitude orbit, which is necessary if it is to be serviced by the Space Shuttle, imposes many constraints. For example, it must always be pointed at least 50 degrees away from the sun, 15 degrees away from the sunlit moon, and 5 to 15 degrees away from the sunlit earth. Add to constraints such as these the over 500 different configurations and combinations of instruments from which an observer can choose, the tasks of correcting the data for instrumental effects and storing it so that it can be readily retrieved, and the demanding and delicate problem of deciding who will get to use the telescope, and you can begin to understand the magnitude of the task of coordinating the scientific activities of the Space Telescope.

In the early planning stages, it was assumed that the science operations would be conducted at Goddard Space Flight Center. Goddard would have responsibility for the development of the scientific instruments before launch and the communications with the spacecraft after launch; therefore, it seemed natural, especially to people within NASA, that Goddard would serve as the scientific nerve center for the project. Astronomers outside NASA didn't see it that way. They wanted the science operations to be run by a group that would be independent of the government, in much the same way that Kitt Peak National Observatory, the National Radio Astronomy Observatory, and Brookhaven National Laboratory are run by associations of universities.

Lyman Spitzer, who has been on practically every committee that discussed or planned the Space Telescope from 1962 onward, recalled that one of these committees discussed how the scientific activity should be organized. "It was the general feeling of the group," he said, "that we would be better off with what we were used to in other government astronomical enterprises, that is, something run by a consortium of universities, rather than by the government directly. That's been the pattern in experimental physics and also in astronomy."

O'Dell advocated this point of view from within NASA. "At the time [1976, when they were making plans for the Space Telescope Science Institute], Goddard did not have an enviable record of doing space astronomy. Remember the OAOs [the Orbiting Astronomical Observatories]. Most of the science from that program came from Princeton, because with *Copernicus* [the name given the second successful Orbiting Astronomical Observatory], it was Princeton people that ran it, and Princeton people doing the science. So at the time, Goddard appeared very weak in space optical astronomy. That's one of the strong reasons why we pushed for the separate science institute approach."

The advocates for a separate science institute won. In January of 1981, NASA awarded a contract to establish and manage the Space Telescope Science Institute to the Association of Universities for Research in Astronomy (AURA), a consortium of 17 universities that operates the United States national optical observatories at Kitt Peak in Arizona, Cerro Tololo in Chile, and Sacramento Peak in New Mexico. Johns Hopkins University was chosen as the site for the institute over such places as Princeton primarily because of its location. It was close enough to Goddard Space Flight Center for easy communication, yet not so close that it would lose its independence.

Most astronomers expected O'Dell to be the director of the Institute, since he had been intimately involved with the project for so long. O'Dell was among them; he looked forward to the directorship as a natural consequence of his dedicated and capable service. Yet in the end, it was this service that made it impossible for him to be the director. "The Carter administration passed a law," he explained. ". . . controlling employment after you leave government service." This law forbids former government employees from ever working in the private sector on projects that they helped to create while they were government employees. "For example," he said, "a general could push and push for a new type of field tank, and once the program got started, he could go to work for Chrysler right away and line his pockets carrying out a program that he had created . . . [To prevent that] was the intent of the law, but the wording of the law applied to someone like me. I was creating conceptually within the agency this idea of a space telescope science institute, and then wanting to go to work for it, presumably as the leader, and that was in direct conflict with this new law."

With O'Dell legally unable to serve as the director, AURA approached David Heeschen. He refused. Then Riccardo Giacconi's name came up. "An extraordinarily strong, appropriate candidate" was John Bahcall's reaction. Others agreed, and Giacconi was offered the position. He was a professor at Harvard and an associate director of the Harvard-Smithsonian Center for Astrophysics. He had a beautiful house within walking distance of work and a seaside home on Cape Cod. The *Einstein* observatory had ceased operation, and although analysis of the data would continue for years, it was not the same as the hectic, intoxicating immediacy of building or operating a telescope in space. For the first time in his career, Giacconi had the leisure to slip into the comfortable role of a Harvard professor, to indulge in his hobbies of painting and sailing, and to just relax after years that Mirella Giacconi termed "breathless. Hurry, hurry through life . . . we built houses, bought houses, sold houses, raised children, dogs, worked in the garden, put in a wall here, planted a tree and left it there. It was very much a rushing through life." Giacconi tried the quiet life for a while, but it didn't fit. He felt miscast. He wasn't a professor; he was a scientific manager.

One of Giacconi's dreams while at American Science & Engineering had been to establish a national x-ray astronomy institute, after the pattern of the National Radio Astronomy Observatory. "It was," he said, "one of the motivating reasons to go from AS&E to Harvard, so as to have the appropriate setting from which you could do such a thing." While at the Harvard-Smithsonian Observatory, he made some progress toward this goal, but it was slow and uncertain. The Space Telescope Science Institute was, he said, "something right along the line of what I wanted to do." So he accepted the offer, and in September 1981 became director of the Institute.

Giacconi saw the chance to apply the management approach that had been so successful at AS&E, without the limitations inherent in a corporation doing business for profit. He had done it within the limits of the Space Science Division at AS&E until company politics had interfered, and he had done it at the Harvard-Smithsonian Observatory within the limits of the High Energy Astrophysics Division. But he was still bound by Observatory politics and procedures, and his prospects for becoming director of the Observatory were remote. "Here, finally," he explained, "I had a chance to do the thing as I thought it should be done. And in that sense it is a very interesting sociological experiment. We tried to write down the management scheme, for instance, being very careful not to overdirect science . . . I'm more ambitious about this institute than I ever was about, say, the x-ray astronomy group at Harvard. There, I think we carried out concepts which were developed in the sixties. That essentially was execution. Here, I think that what we are attempting to do is something very different, because we are supposed to be a national facility that serves the community. That we certainly will do. The question at issue is a different one, whether there will be *research* done here. Whether in general we will be a positive force in astronomy. Whether individuals can develop and flourish here and whether there will be an opportunity for great guys to emerge . . . Well, that's a hell of a lot more difficult proposition than trying to do a job."

It is also, apparently, a hell of a lot more than NASA intended. Almost immediately after taking charge, Giacconi sought to increase the scientific staff and facilities. NASA headquarters reacted with alarm, and the charge of empire building to the detriment of the rest of the astonomical community was raised. Giacconi conceded that the increasingly expensive space research projects pose a problem that must be reckoned with, but he refused to shoulder the blame for that. "If the fact of the matter is that *AXAF* [the Advanced X-ray Astrophysics Facility] is going to cost $750 million to a billion, *SIRTF* [Shuttle Infrared Telescope Facility] will cost a billion, and each one of these things will cost $150 million to operate, then quickly we can't do anything in the United States as far as space astronomy except those things. So it is clearly a problem, but it isn't the . . . Institute."

Critics point to the increase in the planned staff and budget for the Institute from 100 scientists and $5 million a year to 250 scientists and $15 million a year as evidence that the growth of the Institute is out of control. Giacconi responds that this is an unfair interpretation, because the size and budget of the Institute were never clearly established. "If there is a job to do, we do it, and we do it more efficiently than anyone else." He cited the modification of the computer system that will be used to control the science operation from the ground. Before the Institute began operations, TRW designed an inexpensive system that suffered from a lack of advice from its potential users. It was designed more for business applications than for scientific use, had inadequate graphics capability, and would have been expensive to operate. One of the major tasks of the staff at the Institute has been to work with TRW to modify this system to meet the needs of the astronomers who will use the Space Telescope.

We asked O'Dell and Spitzer for their opinions on the issue of the growth of the Space Telescope Science Institute. O'Dell said simply, "It is awesomely large," and declined further comment. In discussing the plans that led to the creation of the Institute, he wondered if perhaps it might not have been better to let Goddard run it after all. A few years after the decision was made not to let Goddard handle the science activities of the Space Telescope, the staff there distinguished themselves with the handling of the International Ultraviolet Explorer. This little jewel of a satellite was a joint venture by NASA, the European Space Agency, and the British Science Research Council. It was launched in 1978 and carried a .45-meter (18-inch) telescope into a geosynchronous orbit over the Atlantic, so that it is in continuous contact with both NASA and European ground stations. For seven years now it has been sending back data on interstellar gas, the hot outer layers of stars, and the nuclei of galaxies and has demonstrated, along with the *Copernicus* satellite, the importance of ultraviolet observations.

"I think that if we were starting today, where Goddard's position is a much stronger one, in large part due to the *IUE*," O'Dell said, "that's a decision that might have gone another way; that is, the general community of astronomers would have been satisfied to let the Goddard people administer the science program."

Spitzer wasn't so sure. "I don't think NASA does things any more efficiently than a civilian organization would do it; there are those who feel that it is somewhat less efficient. That is, I'm sure, a matter of debate. The number of people required is dictated really more by the scientific objectives rather than by the organization."

Did he feel that the growth of the Space Telescope Science Institute was consistent with the scientific objectives of the Space Telescope? "I'm certainly concerned about the size of the Space Telescope Science Institute," he replied, "but, as chairman of STIC [the Space Telescope Institute Council, a senior advisory body of scientists that oversees the management of the institute for AURA], I've been in a good position to see what was happening. In fact, our consent was required for all the various steps that they've taken and I've been convinced that the increases were really needed to carry out the scientific objectives that have previously been agreed upon." Speaking as an astronomer who has been an observer and a participant in space science since its beginning, he offered some perspective. "This isn't the first time that a program . . . has turned out to be somewhat more difficult than estimated. It just takes more funds."

Four different boards have investigated the performance of the Institute and have come away with essentially the same conclusion, namely, that what they are doing is necessary and proper. In May of 1985, the report issued by the Space Telescope Science Institute Task Group of the National Academy of Sciences recommended that "the scope of the Institute should be comparable in budget and manpower to other national astronomical facilities." The report went on to say that "as a result of its study, the Task Group has formed a very favorable impression of the STScI in carrying out the functions indentified . . . [for it] as well as making valuable contributions in identifying and undertaking tasks that were not foreseen at the time. In its short time of existence, the STScI has created a scientific staff of high stature."

Nevertheless, many astronomers will continue to cast a jaundiced eye toward the Institute, because of what it represents—a major and irreversible change in the way they will practice their trade.

A preliminary survey of potential users of the Space Telescope indicated that requests for time on the telescope will be about 15 times greater than can be accommodated. Either the allocation for each project will have to be cut severely, or only about one in fifteen applicants will be selected, or, most probably, the astronomers will form teams which will share the time and data. Many of the larger projects, ones which utilize the unique capabilities of the Space Telescope, will be planned by working groups or committees.

Astronomers who have participated in large satellite projects such as the *Einstein* x-ray observatory and the Infrared Astronomical Satellite are used to working in large groups. But for most optical astronomers, teamwork will be a new experience. The Space Telescope will mark the time when the astronomy community, especially the optical astronomers, officially accept the concept of using large groups to attack key problems. Lone astronomers working on their own projects will become an increasingly rare breed.

___ EPILOGUE

The Space Telescope represents a technological watershed in astronomy, the establishment of a powerful permanent observatory in space. It will be followed by the Gamma Ray Observatory, the Advanced X-ray Astrophysics Facility, and the Shuttle Infrared Telescope Facility, all of which will have the capability to be permanently maintained with the use of the Space Shuttle. Radio astronomers will remain earth-bound for the near future, but they too have plans to put a radio telescope in orbit, which would work in concert with the Very Large Array and other telescopes to mimic a radio telescope larger than the earth.

The trend is clear: larger telescopes placed in space, above the obscuring effects of the atmosphere. This is, of course, extremely expensive. It requires government funding, which means that a consensus of sorts has to be reached before a project can go forward. It also means that fewer new instruments will be built; and thus the demand to use the ones that *are* built will intensify. We are also at a sociological watershed in the history of astronomy. The image of the astronomer is changing from that of a solitary person shivering through the night in a lonely vigil at a telescope on some desolate mountaintop. Rather, the astronomers of the future, especially those who lead the way, will be more like those cosmic inquirers we have encountered in this survey of modern astronomy. They are the polar opposite of the independent scientist working in isolation. They work, of necessity, in and with large groups, and they must have many things besides pure science on their minds. They must be deft politicians who are willing and able to work the hustings for grassroots support and the back rooms for crucial committee votes in favor of their projects. They must be salespersons who can convince NASA, the National Science Foundation, congressional committees, and the Executive Branch that their plan deserves funding. They must be tough, skilled managers who can maintain quality control while meeting the demands of schedule and budget. They must have a constitution and a psyche that can tolerate day after day and year after year of long hours, not just in the lab but in committees, review boards, meetings, and lectures. They must acquire a taste for telephones, airports, and endless paperwork.

But through it all they must retain their ability to dream. Without this ability they will lose sight of their goals, and over the course of the ten, fifteen, or twenty years or more that it takes to bring a project to fruition, they will leave in frustration, or become casualties of the often bitter political warfare, or simply become buried alive in bureaucratic detail.

What about the lone wolves and mavericks, those romantic characters who go their own way, down the less traveled path, and sometimes make discoveries that are totally at odds with the conventional wisdom of the day? People such as Grote Reber, who discovered radio waves from the center of the galaxy with a radio telescope he built in his backyard. Are they finished? Will their contribution to astronomy be negligible in the near future? Maverick astronomers will not be the builders of major new observatories—they never have been—but they will still have a chance to pursue their own line of research without submitting to the will of a group. They can do this with ground-based telescopes, which still have major contributions to make. Access is possible through programs for guest observers at the major space observatories, which will set aside time for small projects. Astronomers can also use the archives of data collected by these observatories; the foundation for the modern theory of the evolution of stars, and the first realistic estimates of the size of our galaxy, came from just such work, namely, the study of the photographic plate collection at Harvard around the turn of the century. With video disks, it will be possible to store and retrieve virtually every bit of data that a telescope collects. This information will be made available to any astronomer who wants to use it one year after it is collected.

Another question is whether or not the system that has evolved for doing big programs is the best one. As the projects get larger and more expensive, committees proliferate and responsibility for the projects becomes diffused. The project leaders will inevitably have to become more conservative, since they will have to seek broader agreement before they can act. This has already happened to a certain extent with the Space Telescope. There was no principal investigator in charge; instead, the activities of the dozens of scientists involved was guided by an eighteen-member science working group. The

Space Telescope Science Institute does have a strong director in charge, but his authority is circumscribed by at least seven advisory boards and committees. If this trend continues, it is difficult to see what will prevent the system from developing a collective inertia that will keep it going forever in the same direction, and where seldom will be heard an original word or new idea.

Some mechanism for encouraging new ideas must be built into the system. This becomes increasingly difficult when peer review determines the worth of an idea. Perhaps an element of randomness should be introduced to keep the system vigorous. The best way to do this, we believe, is to reserve funds for a certain number of smaller projects, and to choose the recipients of these funds by a lottery among those who submit qualified proposals. In ancient Greece, elections by lottery were thought to be more equitable, less likely to encourage political dynasties, and more likely to introduce new faces and ideas into the body politic. In a similar way, a lottery for innovative research might ensure the vitality of astronomy.

Another difficulty with the present system is the adversary relationship that seems to develop between NASA and the scientists. Is it necessary? An adversary relationship is a cornerstone of our legal system, but scientific research programs are not meant to be trials, though they often seem to turn out that way. It has been reported that the Japanese astronomy satellites are built by the Institute for Space and Aeronautical Science of the University of Tokyo for roughly half what a comparable satellite costs NASA. Whether this is because much of the cost of technological developments has been borne by previous NASA space projects, or because doing business with the technically excellent aerospace companies is expensive, or because an adversary relationship develops, or whether some other reason is to blame is not clear. Whatever the case, serious consideration should be given to removing NASA from the development of scientific projects altogether; these could be turned over to the National Science Foundation or some similar agency that would oversee the building of the satellites and contract with NASA for putting them into space. The danger of that course of action is that the science simply would not get done, that the money would be appropriated to and spent by NASA for other nonscience projects. Recall Bob O'Dell's comment about the different color of money for NASA and the National Science Foundation.

Because of the rising cost of astronomical research, and the problems that come with rising costs, many astronomers are gloomy about the future of astronomy. John Bahcall is not. "I'm convinced," he asserted, "that big science, good science, will get at least its fair share of national support. Because it's so interesting and, in the end, so useful to the nation. It keeps us in the technological forefront . . . It also excites the young people to go into technical fields . . . I think it's going to get its fair share because the results of fundamental science are going to appeal to and excite average Americans. A lot of people sell Americans short. I think that when you talk to them and to their congressmen and representatives, they're really interested in the way the world is and and want to know about it. And they're willing to spend a little money, a few percent, on a project to support it."

Certainly the last twenty-five years in astronomy have given anyone interested in the way the world is plenty to be excited about. Before 1960, no one knew that the galaxies were awash in a sea of microwave radiation presumably produced by the Big Bang, or that quasars, pulsars, x-ray stars, gamma ray bursters, and black holes existed. Or that interstellar space contained giant clouds of molecules and dust and vast tunnels filled with hot gas, or that galaxies could undergo awesome bursts of star formation. We didn't know whether our solar system with its planets was unique in the galaxy; we still don't know, but we have strong evidence from infrared observations to indicate that the basic process which led to the formation of our solar system may be a common one. The Space Telescope and other instruments of the future will provide more insight into this and other fundamental questions.

If we answer that question, it will only inspire us to pursue the matter further. Are there other planets such as the earth? Is there extraterrestrial life? Is their intelligent life? What do they know about life and the universe that we don't know? That is the nature of exploration. Each discovery leads to new knowledge, better understanding, and more questions, which lead to more exploration and an enhanced awareness of the complexity, the richness, and the beauty of our cosmic milieu.

Our perception of the universe and the need to know what part we play in its drama are fundamental to our intellectual nature. From the ancient myths and legends conceived when man could only study the heavens with unaided eyes to today's cosmological myths and theories derived from the use of sophisticated instruments and the scientific method, we continue to dream of and search for reality.

It is as if we had awakened to find ourselves in a strange land on a foggy night. Until the fog lifted, we could scarcely see beyond our hands and hadn't the vaguest notion of what the countryside looked like. We could only speculate that there was a mountain in front of us, an icy river behind us, and a black panther grousing around in the dark looking for a meal. Before dawn the fog lifts. It is still dark, but by starlight we can see enough to understand that there is indeed an icy river behind us; but the mountain is only a gentle hill, and the panther is not a panther but a cow grazing in a field near us. We are surprised and intrigued by the delicate, lacy silhouette of a small tree several yards in front of us, and in the distance, our attention is drawn by dark shapes that we had not imagined. Is it a circle of stones—or is it an array of radio antennas? What is it? We move forward in the dawn to investigate.

"Life has a limit, knowledge none."

CHUNG-TZU

Bibliography
Index

THE ELECTROMAGNETIC SPECTRUM

TYPE OF RADIATION	WAVELENGTH	PENETRATES ATMOSPHERE
Radio	1 millimeter–10 meter	Yes
(Microwave)	(1–10 centimeters)	(Yes)
Infrared	1 micron–1 millimeter	Only near 1 micron
Visible	300 nanometers–1 micron	Yes
Ultraviolet	10–300 nanometers	No
X-ray	0.01–10 nanometers	No
Gamma ray	less than 0.01 nanometers	No

Note: 1 meter = 1,000 millimeters = 1,000,000 microns = 1,000,000,000 nanometers.

___ BIBLIOGRAPHY

PART ONE: THE VERY LARGE ARRAY

The basic reference for this part is the article written by David Heeschen, "The Very Large Array," in *Telescopes for the 1980s,* ed. G. Burbridge and A. Hewitt (Palo Alto: Annual Reviews Inc., 1981), pp. 1–62. An authoritative discussion of the early history of radio astronomy by one of the pioneers is given in J. S. Hey, *The Radio Universe* (Oxford: Pergamon, 1983). The discovery of the 21-centimeter line is discussed in I. Shklovsky, *Cosmic Radio Waves* (Cambridge: Harvard University Press, 1960), and in T. Ferris, *The Red Limit* (New York: Morrow-Quill Books, 1983). A general discussion of radio astronomy, with beautiful illustrative material from the VLA and other radio telescopes, is found in N. Henbest and M. Marten, *The New Astronomy* (Cambridge: Cambridge University Press, 1983).

Some interesting anecdotes concerning the problem of radio interference are given in R. Kazarian, "Blankets and Hair Clippers Complicate Radio Studies of the Universe," *Mercury,* July/August 1983, p. 122. In our discussion of the political maneuverings that led to the decision to build the VLA we relied on, in addition to our interviews, the unpublished study by G. Lubkin, "The Decision to Build the Very Large Array" (1975).

For more detailed discussion of some of the problems attacked by VLA, see: J. Burns and R. Marcus, "Centaurus A: The Nearest Active Galaxy," *Scientific American,* November 1983, p. 56; R. Blandford, M. Begelman, and M. Rees, "Cosmic Jets," *Scientific American,* May 1982, p. 124; L. Blitz, "Giant Molecular Cloud Complexes in the Galaxy," *Scientific American,* April 1982, p. 84; C. Lada, "Energetic Outflows from Young Stars," *Scientific American,* July 1982, p. 82; F. Seward, P. Gorenstein, and W. Tucker, "Young Supernova Remnants," *Scientific American,* August 1985, pp. 88–96; and M. K. Crawford et al., "Mass Distribution in the Galactic Centre," *Nature,* 315 (6 June 1985): 467.

PART TWO: THE EINSTEIN X-RAY OBSERVATORY

The basic references for this part are W. Tucker and R. Giacconi, *The X-Ray Universe* (Cambridge: Harvard University Press, 1985), and W. Tucker, *The Star Splitters* (Washington, D.C.: U.S. Government Printing Office, 1984). The early history of x-ray astronomy up through the *Uhuru* satellite, including the discovery that x-ray stars are neutron stars and black holes, is also discussed in R. Hirsh, *Glimpsing an Invisible Universe* (Cambridge: Cambridge University Press, 1983), and in G. Greenstein, *Frozen Star* (New York: Freundlich, 1983).

PART THREE: GAMMA RAY ASTRONOMY

The *SAS-2* and *COS-B* gamma ray experiments are discussed briefly in N. Henbest and M. Marten, *The New Astronomy*, (Cambridge: Cambridge University Press, 1983), pp. 202–205. Trevor Weekes gives an overview of gamma ray astronomy in *Mercury*, May–June 1981, pp. 78–84. NASA's High Energy Astronomy Observatory program is the subject of W. Tucker, *The Star Splitters* (Washington, D.C.: U.S. Government Printing Office, 1984). The origins of gamma ray astronomy are discussed in G. Clark, "Gamma Ray Astronomy," *Scientific American*, May 1962, pp. 52–61; Gamma ray lines are discussed by Marvin Leventhal and Crawford McCallum in "Gamma Ray Line Astronomy," *Scientific American*, July 1980, pp. 62–70; gamma ray bursters are discussed in Bradley Schaefer, "Gamma Ray Bursters," *Scientific American*, February 1985, pp. 52–58. The importance of the aluminum 26 gamma ray line was stressed by W. Fowler in his Nobel Prize lecture, "The Quest for the Origin of the Elements," which is reproduced in *Science*, 226 (1984): 922–935. More technical discussions of various aspects of gamma ray astronomy can be found in the following articles in the *Annual Review of Astronomy and Astrophysics:* L. Peterson, "Instrumental Techniques in X-ray Astronomy," 14 (1975): 423–509; G. Steigman, "Observational Tests of Antimatter Cosmologies," 14 (1976): 339–372; G. Bignami and W. Hermsen, "Galactic Gamma Ray Sources," 21 (1983): 67–108.

PART FOUR: THE INFRARED ASTRONOMICAL SATELLITE

A good introduction to infrared astronomy before the Infrared Astronomical Satellite project is given in N. Henbest and M. Marten, *The New Astronomy* (Cambridge: Cambridge University Press, 1983). The article by G. Neugebauer and E. Becklin, "The Brightest Infrared Sources," in *New Frontiers in Astronomy*, (San Francisco: W. H. Freeman and Co., 1975), pp. 186–199, summarizes the state of infrared astronomy in 1973. For more detail, see the book by D. Allen, *Infrared, the New Astronomy* (New York: Wiley, 1975).

A summary of the Infrared Astronomical Satellite project is given by H. Habing and G. Neugebauer, the co-chairmen of the *IRAS* science team, in "The Infrared Sky," *Scientific American*, November 1984, pp. 49–57. Summaries also appear in G. Neugebauer et al., "Early Results from the Infrared Astronomical Satellite," *Science*, 224 (April 6, 1984): 13–21; and "IRAS, the Infrared Astronomical Satellite," *Nature*, 303 (May 26, 1983): 287–291. MOSFETS and JFETS are discussed in G. Miller, *Modern Electricity/Electronics* (Englewood Cliffs, N.J.: Prentice-Hall, Inc., 1981). The difficulties encountered in building *IRAS* are discussed from the engineer's point of view in an article by M. Waldrop, "The Infrared Astronomy Satellite (I)," *Science*, 220 (24 June 1983): 1363–1368. Mission operations at the Rutherford Appleton Laboratory in England are discussed by J. Macdougall et al., "IRAS Mission Operations Experience at RAL," *Journal of the British Interplanetary Society*, 37 (1984): 337–347. The data-processing problems and the early results from *IRAS* are discussed by M. Waldrop in "The Infrared Astronomy Satellite (II)," 221 (1 July 1983): 43–45.

The quotations from David Allen are from D. Allen, "Infrared Astronomy: An Assessment," *Quarterly Journal of the Royal Astronomical Society*, 18 (1977): 188–198. The quotation from John Duxbury is from M. Waldrop's article in the 1 July 1983 issue of *Science*. The fast-moving object search is described by J. Davies et al. in "The IRAS Fast-moving Object Search," *Nature*, 309 (24 May 1984): 315–319. The Spacewatch project is described in T. Gehrels, "Asteroids and Comets," *Physics Today*, February 1985, pp. 33–41.

The 1 March 1984 issue of *Astrophysical Journal* contains 21 articles that summarize the mission and give the first published results from *IRAS*, including those on the comet IRAS-Araki-Alcock, the zodiacal bands, and the shell around Vega. The discovery of the intense infrared radiation from Arp 220 is reported by B. Soifer et al. in *The Astrophysical Journal*, 283 (1984): L1–L4.

Technical reviews of infrared astronomy are given in the following *Annual Reviews of Astronomy and Astrophysics* articles: H. Spinrad and R. Wing, "Infrared Spectra of Stars," 7 (1969): 249–302; G. Neugebauer, E. Becklin, and A. Hyland, "Infrared Sources of Radiation," 9 (1971): 67–102; B. T. Soifer and J. L. Pipher, "Instrumentation for Infrared Astronomy," 16 (1978): 335–369; G. Rieke and M. Lebofsky, "Infrared Emission of Extragalactic Sources" 17: 477–511; K. Merrill and S. Ridgway, "Infrared Spectroscopy of Stars," 17 (1979): 9–41; S. Kleinmann, F. Gillett, and R. Joyce, "Preliminary Results of the Air Force Infrared Sky Survey," 19 (1981): 411–456; C. Wynn-Williams, "The Search for Infrared Protostars," 20 (1982): 587–618; W. Stein and B. Soifer, "Dust in Galaxies," 21 (1983): 177–207.

PART FIVE: THE SPACE TELESCOPE

A good overview of the pre–Space-Telescope status of optical astronomy can be found in N. Henbest and M. Marten, *The New Astronomy* (Cambridge: Cambridge University Press, 1983), pp. 40–55. The early history of rocketry, including a brief discussion of project RAND, is given in A. C. Clarke, *Man and Space* (New York: Time, Inc., 1964). The limitations of ground-based telescopes are discussed in Henbest and Marten (1983), pp. 53–55, and in C. R. O'Dell, "The Space Telescope," *Telescopes for the 1980s* (Palo Alto: Annual Reviews, Inc., 1981), pp. 129–194. The article by O'Dell is a basic reference that discusses the early history of the Space Telescope, the design considerations, the politics, and the scientific instruments in some detail. Another overview of the project is given by John Bahcall and Lyman Spitzer, Jr., in "The Space Telescope," *Scientific American*, July 1982, pp. 40–52. A more technical, and very detailed, description of the Space Telescope is given in the NASA publication, *The Space Telescope Observatory*, ed. D. Hall (Washington, D.C.: NASA, 1982), which is scheduled to be republished by the Sky Publishing Company.

Ultraviolet astronomy and the *Copernicus* and International Ultraviolet Explorer satellites are discussed in Henbest and Marten (1983), pp. 152–159, and in L. Spitzer, *Searching between the Stars* (New Haven: Yale University Press, 1982).

A description of Spitzer's and Bahcall's lobbying activities and experiences is given in J. Bahcall, "Galaxies, Quasars and Beyond: The Space Telescope," *Space Applications at the Crossroads, 21st Goddard Memorial Symposium*, Vol. 55, Science and Technology Series, ed. J. McElroy and E. Heacock (San Diego: Univelt, 1983); and in P. Hanle, "Astronomers, Congress and the Large Space Telescope," *Sky and Telescope*, April 1985, pp. 300–305, wherein the quotation from the Congressional Appropriations Committee is given. This article also has a sidebar by K. Beatty on the contribution of technology developed by the military to the Space Telescope project.

The problems encounterd in building the Space Telescope are discussed in "Washington Reports," *Physics Today*, November 1983, pp. 47–49, wherein the quotation from J. Beggs is given; in M. Waldrop, "Space Telescope in Trouble," *Science*, 220 (8 April 1983): 172–174; in J. K. Beatty, "Space Telescope: Problems and Progress," *Sky and Telescope*, September 1983, pp. 189–190; in "Search and Discovery," *Physics Today*, November 1984, pp. 17–19.

The coating of the mirror is discussed in L. Robinson, "An Eye for Tomorrow," *Sky and Telescope*, February 1982, p. 128.

The Space Telescope Science Institute is discussed in M. Waldrop, "Space Telescope (II): A Science Institute," *Science*, 221 (5 August 1983): 534–536; and in W. Tucker, "The Space Telescope Science Institute," *Sky and Telescope*, April 1985, pp. 295–299.

An overview of some of the scientific problems that will be attacked by the Space Telescope is given in M. Longhair, "The Scientific Challenge of the Space Telescope," *Sky and Telescope*, April 1985, pp. 306–311.

ADDITIONAL READING

Up-to-date reports on new discoveries in astronomy can be found in the following nontechnical magazines: *Astronomy, Mercury* (a journal of the Astronomical Society of the Pacific), *Scientific American*, and *Sky and Telescope*.

INDEX